Essential Physics for Radiographers

Essential Physics for Radiographers

J. L. BALL and A. D. MOORE
TDCR TDCR

South West Wales
School of Radiography
Morriston Hospital
Swansea

FOREWORD BY

MURIEL O. CHESNEY
FSR TE

Teacher Principal
United Birmingham Hospitals
School of Radiography
The General Hospital
Birmingham

Blackwell Scientific Publications

OXFORD LONDON EDINBURGH
BOSTON MELBOURNE

Editorial offices:
Osney Mead, Oxford, OX2 0EL
8 John Street, London, WC1N 2ES
9 Forrest Road, Edinburgh, EH1 2QH
52 Beacon Street, Boston,
Mass., USA
214 Berkeley Street, Carlton
Victoria 3053, Australia

First published 1980

Set by Santype International Ltd.
Salisbury, Wiltshire.
Printed and bound in Great Britain by
Billing and Sons Ltd.
Guildford, London and Worcester

DISTRIBUTORS

USA
Blackwell Mosby Book Distributors
11830 Westline Industrial Drive,
St Louis, Missouri 63141

Canada
Blackwell Mosby Book Distributors
86 Northline Road, Toronto
Ontario, M4B 3E5

Australia
Blackwell Scientific Book
Distributors
214 Berkeley Street, Carlton
Victoria 3053

British Library
Cataloguing in Publication Data

Ball, J L
Essential physics for radiographers.
1. Radiography
2. Medical physics
I. Title
530'.02'4616 RC78

ISBN 0-632-00644-7

Contents

Foreword

It has been said that physics is a subject which student radiographers find difficult. I sometimes think that while it certainly used to be true, perhaps the notion is now just beginning to acquire the status of a myth. Be that as it may, what is certainly true is that there is a need for physics to be taught to radiographers by radiographers—not by those who understand a vast quantity of physics but have no conception of the deep well of miscomprehension into which they may throw their knowledge when they seek to pass it on in schools of radiography.

So a book on the physics required for the Diploma of the College of Radiographers which has been written by radiographers has in principle my full support. In particular, for one written by these two radiographer/teachers I come with enthusiasm to the task of sending it forward (if you'll forgive me!) on its way. I know the authors well and my knowledge of their experience and abilities makes me sure that they are the right men for the task that they have assumed. Their work should help to make coming generations of student radiographers wonder why their predecessors found physics so high a hurdle to clear with success.

Birmingham 1980 *Muriel O. Chesney*

Preface

Physics for the Diploma of the College of Radiographers' examination is a subject with which many students experience difficulty. With this in mind we feel there is a need for a book written *by* radiographers *for* radiographers. Our aim has been to provide a clear and easily understandable text which the reader can use without becoming bored or 'bogged down'.

We have tailored the physics specifically to the needs of the student radiographer by relating it to everyday experiences, both at home and at work. The mathematics has been kept to a minimum and is accompanied by full and clear explanations. No new terms are introduced without first being explained. To simplify in these ways has sometimes meant telling less than the whole truth, in order to achieve a clear understanding.

The book is divided into twenty chapters which correspond to the physics course lectures at the South West Wales School of Radiography. Although this book is primarily intended for Part I DCR students, it will also be found useful prereading for candidates of the higher diploma.

Swansea 1980

J. L. Ball
A. D. Moore

Chapter 1
General physics

MATTER AND ENERGY

Physics is concerned with the study of two concepts: matter and energy, and the relationships between them.

MATTER

Matter is the name given to the material of which everything is made. It normally exists in one or more of the three physical states: solid, liquid, gas.

In general, it is not possible to destroy matter, and it is not possible to create it out of nothing. However, it can be converted from one form to another by physical or chemical means, for example:

1. Ice can be melted and turned from a solid into a liquid (water). This is a *physical* change because ice and water are two forms of the same substance.
2. Wood can be burnt and changed into ashes and smoke. This is a *chemical* change since wood and ash are not the same substance.

Physicists summarise this idea by saying that 'matter can neither be created nor destroyed', and they call this the *Law of Conservation of Matter*. (In physics, a law is a statement which has been proved by experiment or observation to be universally true.)

ENERGY

Energy is described as the ability to do work. This can be explained by considering what would happen if there were no energy available. Without energy nothing would happen, nothing would ever change, nothing would ever get done. Energy is needed to make things happen. It exists in many forms and can be transformed from one form to another, as shown by the following examples:

1. By rubbing the hands together, energy of movement (kinetic energy) is transformed into heat energy. This transformation is easy to achieve.
2. Transforming heat energy into movement can also be done, as in a steam engine; but this transformation is not efficient and a lot of the heat energy is not converted at all.

Again, as with matter, it is not generally possible to create or destroy energy; only to transform it from one type to another. Physicists call this the *Law of Conservation of Energy*.

Since early this century, it has been known that matter and energy are really two forms of the same thing, and that one can be changed into the other; i.e. matter *can* be created (from energy), and energy *can* be created (from matter), as in a nuclear reactor or atomic bomb. However, the total amount of energy and matter is constant in a self-contained system (the Law of Conservation of Matter and Energy).

Einstein's famous equation, $E = mc^2$, enables us to work out how much energy is produced when matter is converted into energy, and also to determine how much matter is produced from energy. (E = energy; m = mass; c = a constant, the speed of light. A constant is a number having a fixed value which is used in science.)

MEASUREMENT AND UNITS

Physics is concerned with quantities. As well as knowing what and why things happen, physicists also like to know *how much*; i.e. measurements are essential in physics.

There are three fundamental quantities which can be measured; all other quantities are derived from them. We call the basic quantities 'dimensions'. They are: mass (M)—the quantity of matter; length (L); and time (T). The derived quantities are combinations of the basic dimensions, e.g. area is L^2; volume is L^3; speed is L/T (distance covered divided by time taken); density is M/L^3 (mass divided by volume); acceleration is L/T^2 (increase in speed divided by time taken); and so on.

In order for measurements to have meaning, units of measurement have to be created and be widely accepted among the people who are going to make and use the measurements. Several systems of units are available, for example:

1. Imperial system (British). The basic units are the pound (mass), the foot (length), and the second (time). It has been in common use in everyday life in Britain for hundreds of years and is now gradually being phased out.
2. Continental system (c.g.s.). The basic units are the centimetre (length), the gram (mass), and the second (time). It has traditionally been used in science and is still used in many older textbooks and by the older generation of scientists.
3. Système International (SI). The basic units are the metre (length), the kilogram (mass), and the second (time). It is now widely used in science and engineering and is slowly gaining more general use in the United Kingdom.

We shall use SI units throughout this book as our basic measuring system.

BASIC AND DERIVED UNITS

As we have noted above, three basic SI units are in use: the metre, the kilogram and the second. The metre (m) is equivalent to a length of about 39 inches. The kilogram (kg) is equivalent to about 2.2 pounds. The second (s) is 1/86 400th of a day.

The basic units have multiple and fractional sub-units;

e.g. 1000 metres = 1 kilometre (km)
1/100th m = 1 centimetre (cm)
1/1000th m = 1 millimetre (mm)
1/1 000 000th m = 1 micrometre or micron (abbreviated 'μm' or just 'μ', the Greek letter 'mu')
1/1 000 000 000th m = 1 nanometre (nm)
1 kilogram = 100 grams (g)
1/1000th g = 1 milligram (mg)
1/1000th second = 1 millisecond (ms)
1/1 000 000th second = 1 microsecond (μs).

Physical quantities other than mass, length or time are measured in *derived* units which are combinations of the basic units; e.g. area is measured in square metres (m^2); speed is measured in metres per second (m/s or m s^{-1}); and density is measured in kilograms per cubic metre (kg/m^3 or kg m^{-3}).

Some derived units are given special names for the sake of simplicity; e.g. energy units are kg m^2/s^2 and are called *joules*; volume units are m^3 in SI, although the cubic centimetre (cm^3) is still widely used. One litre (l) is 1000 cm^3 and the cubic centimetre is known as the millilitre (ml).

PHYSICAL QUANTITIES

In physics many commonplace terms have very specialised meanings. We shall consider three of the most important quantities: force, work and energy.

FORCE

Force is defined as that which moves or tends to move a stationary body. It also speeds up or slows down a moving body. In other words it is a 'push' or a 'pull'.

The unit of force is a derived unit resulting from the product of mass and acceleration (kg × m/s^2)—this is called the *newton*, (N).

Definition. A force of 1 newton will produce an acceleration of 1 metre per second squared (1 m/s^2) in a body whose mass is 1 kilogram, if it is free to move and not acted upon by any other forces.

When describing a force it is not enough to quote the magnitude (i.e. strength) of the force; e.g. 6 N. It is also necessary to specify its direction; e.g. vertically downwards. Quantities such as this, having direction as well as magnitude are known as *vector* quantities, or *vectors*. Quantities having only magnitude are known as *scalar* quantities; e.g. mass. A useful example of the difference between these concepts is the case of *weight* (which is a force and thus a *vector*) and *mass* (which is scalar with no direction implied). In everyday life we use the same units for both mass and weight (mass units such as kilograms or pounds). In science we should use force units (newtons) for weight and mass units (kilograms) for mass. The relationship between the two units is relatively straightforward: newtons = kilograms × g (g is the acceleration of a body when falling under gravity and is 9.81 m/s^2).

So 1 newton = 9.81 kilograms (or roughly 10 kg).

WORK

If a force succeeds in producing or stopping movement we say that *work* has been done and we define work done as the product of the magnitude of the force and the distance moved;

i.e. work done = force × distance moved
(in the direction of the force).

The units in which work is measured are newton metres (i.e. newtons × metres) and are renamed *joules* (J). (1 joule = 1 newton metre.)

The scientific meaning of the term 'work' can produce consequences which seem at odds with our everyday experiences. Suppose a patient has collapsed on the floor and we try to lift him on to a stretcher trolley. If we succeed in lifting him up we have done work (we have applied a force, the lift, and moved the patient through a certain distance upwards in the direction of the force).

So far, so good. But what if we were unable to lift the unfortunate patient because of his weight? The physicist would say we had done *no* work on the patient because we had not moved him. However, we may well feel that we tried very hard to achieve the lift and even though our efforts were in vain we had definitely been working!

Work can be thought of as being energy *usefully* expended, and since, in our example, we did not move the patient in the direction of our applied force, any energy we used was 'wasted'.

ENERGY

As we saw at the beginning of this chapter, energy is defined as the ability to do work.

Its unit of measurement is the same as the unit of work; i.e. the joule. In fact, the joule is a rather small unit and we shall meet larger units of energy later in the course.

Energy can appear in many forms, for example:

1. Mechanical potential energy is the energy stored in a body by virtue of its position; e.g. in the case of a compressed spring, or a sandbag on a high shelf. In each case when the object is released it gives up its stored potential energy. Potential energy is often abbreviated 'PE'.

2. Kinetic energy is the energy a body has by virtue of its movement; e.g. a falling sandbag and a moving stretcher trolley both possess kinetic energy. Kinetic energy is abbreviated 'KE'.

The potential energy a body has is equal to the amount of work that was done in putting the body in its particular position; i.e. the force applied multiplied by the distance moved. In the case of gravitational potential energy,

$$\mathrm{PE} = mgh$$

where 'm' is mass, 'g' is the acceleration due to gravity and 'h' is the height the body is raised. The kinetic energy of a body is given by:

$$\mathrm{KE} = \frac{1}{2}mv^2$$

where v is the speed at which the body is travelling.

There are many other forms of energy in addition to the two we have described above; e.g. chemical energy, heat energy, electrical energy, nuclear energy, sound energy, x-ray energy, etc. We shall be looking at some of these in later chapters.

It may be important on many occasions to know the *rate* at which we are expending energy in a particular process or action. We call the rate of using energy 'power'. It can also be defined as the rate of doing work, i.e.

$$\text{Power} = \frac{\text{work done}}{\text{time taken}}.$$

The unit of power will therefore be joules per second (J/s). This derived unit is also known as the watt (W); i.e. 1 watt = 1 joule per second. The watt is a relatively small unit being hundreds of times smaller than the older, more well known unit, the horse power (1 horse power = 746 watts).

The ideas introduced in this chapter will be referred to many times in later chapters. If possible, talk over the work with your tutors and with your fellow students until you are confident that you understand it. Some simple problems follow which should help you achieve a proper understanding. The answers are given in brackets after each question.

PROBLEMS

1. An 18-month-old baby has a mass of 10 kg (22 lb). What is its weight in newtons? (Assume '*g*' is 9.81 m/s^2.) *(98.1 N)*

2. 1 ml of water has a mass of 1 g. What is the weight of 1 litre of water? *(9.81 N)*

3. How fast will a body be travelling after it has fallen under gravity for 10 seconds? (Neglect air resistance.) *(98.1 m/s)*

4. How much work is done in lifting a 100-kg patient from the floor on to a stretcher trolley 1 metre high? *(981 J)*

5. How much power would be required to achieve this lift in 1 second? *(981 W)*

6. How much energy would be released if the same patient fell off the stretcher trolley on to the floor? *(981 J)*

7. At what speed would he hit the floor? (*Hint*: think about kinetic and potential energy.) *(4.3 m/s approx.)*

8. What force would be required to accelerate a stretcher trolley at 1 m/s^2 if its mass is 20 kg and it carries a patient of 80 kg? *(100 N)*

9. How fast will it be travelling if this force is maintained for 5 seconds? *(5 m/s)*

10. How much energy would be needed to bring the trolley to rest? *(1250 J)*

CHAPTER SUMMARY

1. Physics is the study of the relationship between matter and energy (p. 1).

2. Matter is the material of which everything is made (p. 1).

3. Energy is the ability to do work (p. 1).

4. Neither energy nor matter can be created or destroyed, but one can be converted into the other (p. 2).

5. The three basic dimensions of measurement are mass, length and time (p. 2).

6. The SI system of units is based on the kilogram, the metre and the second (p. 3).

7. Force is that which moves or tends to move a stationary body. It is measured in newtons (p. 4).

8. Energy appears in many forms. It is measured in joules (pp. 5 & 6).
9. Power is the rate of doing work. It is measured in watts (p. 6).

KEY RELATIONSHIPS
Work (and energy) = force × distance (p. 5).
Potential energy = mgh (p. 6).
Kinetic energy = $\frac{1}{2}mv^2$ (p. 6).
1 watt = 1 joule per second (p. 6).

Chapter 2
Heat

All matter is made up of minute particles, too small to be seen even with a good microscope. The particles are called *atoms* and *molecules* and we shall consider them in detail in Chapter 4. The important property of these particles that we are concerned with in this Chapter is the fact that they are always in constant motion. The type of motion depends on whether the material is a solid, liquid or gas (see Fig. 2/1).

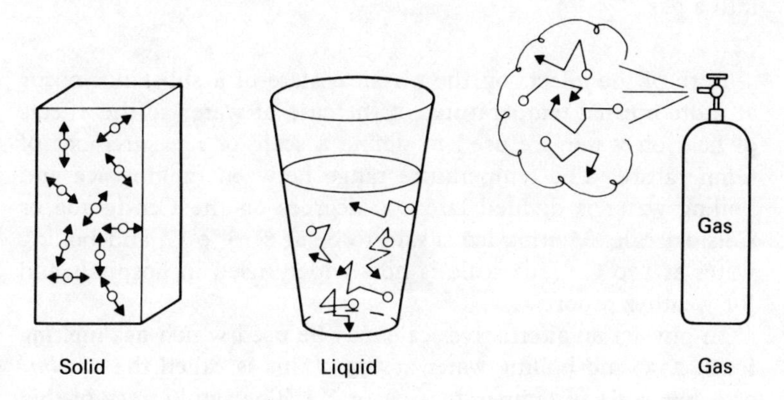

Fig. 2/1. Motion of particles in solids, liquids and gases.

In solids the particles vibrate about fairly fixed positions. This is the reason why solids do not alter their shape easily.

In liquids and gases the particles move randomly. Liquids and gases do not have fixed shapes for this reason.

Because these particles are in constant motion, they possess *kinetic energy*. We call the kinetic energy of atoms and molecules *heat*. So heat is a form of energy as we indicated in Chapter 1.

TEMPERATURE

Temperature is an indication of the average speed of movement of the particles. A high temperature means that the particles are moving very quickly, and a low temperature means they are moving relatively slowly. Temperature changes cause various effects; for example:

1. If temperature is increased, a substance will expand; if it is reduced the substance will contract. Mercury thermometers work on the principle of expansion and contraction. If expansion or contraction is prevented, then a change of pressure will be experienced.
2. If the temperature of a solid is increased it will eventually melt and change from the solid into the liquid state.
3. If the temperature of a liquid is increased it will boil and turn into a gas.

Both of the effects on the physical state of a substance occur at quite normal temperatures in the case of water so the effects of heat on water are used to define a scale of measurement of temperature. The temperature range between melting ice and boiling water is divided into 100 degrees on the Centigrade or Celsius scale. Melting ice is said to be at zero (0°C) and boiling water at 100°C. This scale is now widely used in hospitals and for weather reports.

In physics an alternative scale may be used which has melting ice at 273° and boiling water at 373°. This is called the *absolute* or *kelvin* scale of temperature (A or K). The significance of this scale is that at zero (0 K) the atomic particles which are normally in motion are at rest. In other words, at 0 K a body has no heat energy at all. This temperature, which cannot be reached except in the imagination, is known as *absolute zero*. Scientists have approached this temperature very closely but have not been able to bring those last few particles to rest (see Fig. 2/2).

The Fahrenheit scale divides the range between melting ice and boiling water into 180 degrees, melting ice being at 32°F and boiling water at 212°F. Body temperature is about 98.4°F, or 37°C, or 310 K.

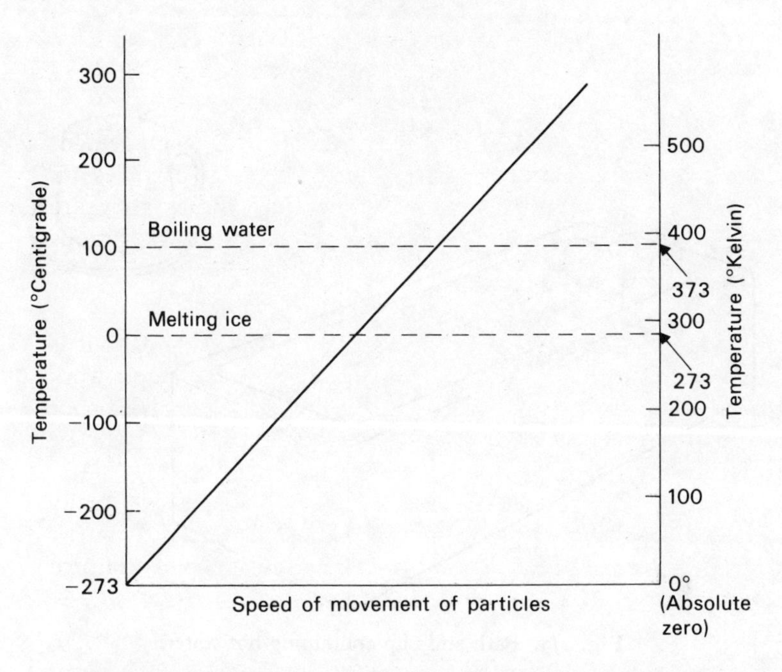

Fig. 2/2. Temperature scales related to particle movement.

HEAT ENERGY

It is important to make clear the difference between heat and temperature. Heat is the total *energy* of motion of the atoms and molecules in a material. Temperature is their average *speed* of motion.

Heat *flows* from hot to cold bodies. At a particular temperature a large body will possess more heat than a small one of the same material; e.g. in Fig. 2/3 the bath full of water contains more heat energy than the cup of water even though they are both at the same temperature.

Heat is a very common form for energy to take. Most energy will, in time, be converted into heat energy and as we shall see in later chapters, much of the energy used by x-ray equipment is transformed into unwanted heat.

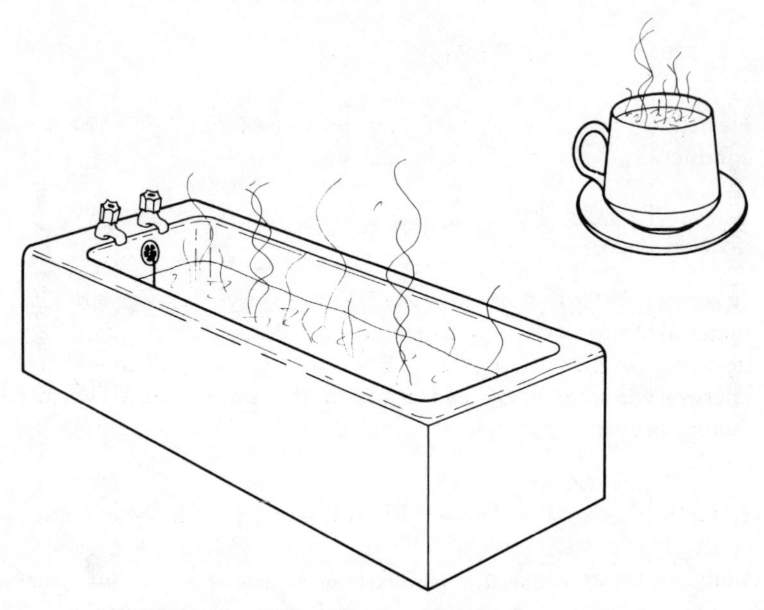

Fig. 2/3. Bath and cup containing hot water.

UNITS OF HEAT

The SI unit of heat is the *joule* because heat is a form of energy. However, a commonly used alternative unit is the *calorie*, defined as being the amount of heat required to cause a 1°C rise in temperature in 1 gram of water. One calorie is equivalent to about 4.2 joules. One gram of most other substances requires less heat than water to raise its temperature by 1°C. The amount is known as the *specific heat* and it is related to heat required by:

$$\text{Heat} = \text{mass} \times \text{specific heat} \times \text{temperature rise}.$$

The specific heat of water is 1 calorie/gram °C.
The specific heat of copper is 0.1 calorie/gram °C.
We sometimes want to know the heat required to bring about a 1°C temperature rise in a component, such as part of an x-ray tube, which does not weigh 1 gram. We call this the *thermal capacity* of the component. In an x-ray tube, the part called the anode, which has a lot of heat to absorb, is made with a high thermal capacity so that it can accept large quantities of heat without its temperature rising to a dangerous level.

MOVEMENT OF HEAT

Heat flows from one place to another in different ways; namely conduction, convection, and radiation.

CONDUCTION

In solids heat flows by conduction. This is the passage through the material of the atomic vibrations which heat produces by collisions between neighbouring atoms. In this way the vibrations (heat energy) are able to spread through the entire solid. Various factors influence the rate at which heat is conducted through a solid:

1. Cross-sectional area. A thick solid can transmit heat more easily than a thin one (heat finds it more difficult to 'squeeze' along a thin conductor):

$$\text{Heat flow} \propto \text{cross-sectional area } (A)$$

where $\propto$ means 'is proportional to'.
2. Temperature difference. If there is a big difference in temperature between opposite ends of the solid, heat will flow more quickly.

$$\text{Heat flow} \propto \text{temperature difference } (T_2 - T_1).$$

3. Length (the distance the heat has to travel). The greater the distance the slower the heat flow:

$$\text{Heat flow} \propto \frac{1}{\text{length of path } (l)}.$$

4. Material involved. Some materials are better at conducting heat than others, e.g. metals are better than plastics. We call the conducting property the *thermal conductivity* (k):

$$\text{Heat flow} \propto k.$$

Summarising the factors affecting rate of heat flow we get:

$$\text{Rate of heat flow} = kA(T_2 - T_1)/l \text{ calories per second.}$$

CONVECTION

In liquids and gases some conduction takes place but heat flow is mainly by convection. This is the movement of heat energy by circulation of the heated liquid or gas, forming convection currents. When a liquid (or gas) is heated it becomes less dense (lighter) and rises to be replaced by cooler liquid (or gas). This is the reason why convection currents are set up. Fig. 2/4

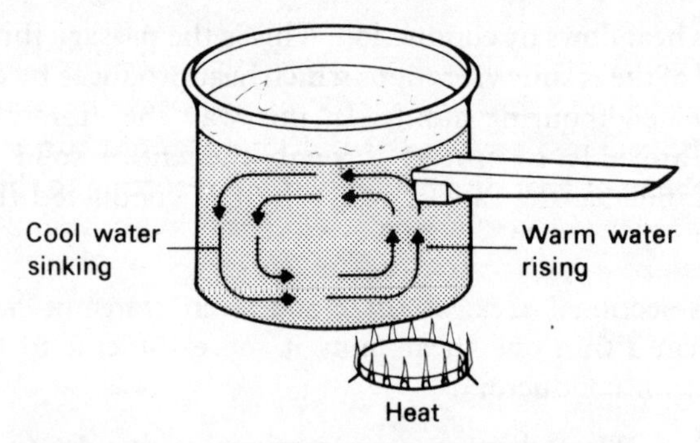

Fig. 2/4. Convection currents in heated liquid.

illustrates this process. Convection currents are used in x-ray tubes to help disperse the heat produced when the tube is operating and we will be discussing this aspect of x-ray tube design in Chapter 12.

RADIATION

In a vacuum (e.g. in space) heat can flow from one place to another even though there is no solid, liquid or gas in between. The process by which heat energy is transmitted through a vacuum is called radiation. It is the method by which heat reaches us from the sun. When atoms and molecules vibrate they emit rays which consist of pure energy. They are known as electromagnetic waves.

Every body emits heat rays of this type because in all bodies the atoms and molecules are in constant motion (only at absolute zero does this movement cease). The amount of heat energy radiated depends on the temperature of the body: energy is proportional to T^4, where T is the temperature in K. So if the temperature of a body is doubled, e.g. from 200 K to 400 K, sixteen times more heat energy would be radiated. A body may also *absorb* heat energy from an external source and at the same time be radiating heat energy. In such a situation, if the body absorbs more heat than it radiates its temperature will rise. If it radiates more than it absorbs it will become cooler and its temperature will fall. If the heat absorbed balances the heat radiated, the temperature will remain constant.

The *surface colour* and *surface texture* of a body also affect the quantity of heat radiated or absorbed. A black and matt surface will radiate and absorb heat efficiently, while a white and glossy surface will not. Until recently x-ray tubes were encased in metal housings which had a black, dull finish so that the internal heat could be radiated away more quickly. Modern tubes overcome the problem of heat dissipation without resorting to this method.

In later chapters we shall consider how these various methods of transferring heat from place to place are used in overcoming heating problems in x-ray equipment.

CHAPTER SUMMARY

1. All matter is made of atoms and molecules (p. 9).
2. Heat is the energy of motion of atoms and molecules (p. 9).
3. Temperature is the speed of movement of atoms and molecules (p. 10).
4. The boiling point of water is 100°C and the melting point of ice is 0°C (p. 10).
5. Heat energy may be measured in joules or calories (p. 12).
6. Specific heat is the heat required to raise the temperature of 1 g of a substance by 1°C (p. 12).
7. A body with a high thermal capacity can accept large amounts of heat with only a small rise in temperature (p. 12).
8. Heat flows by conduction through solids (p. 13).
9. Rate of heat flow by conduction depends on the conductivity of the material (p. 13).
10. Heat flows by convection through liquids and gases (p. 14).
11. Heat is radiated through a vacuum (p. 14).
12. The surface finish of a body influences the quantity of heat it radiates (p. 15).

Chapter 3
Electricity

In this chapter we discuss some of the basic properties of electric charges. This will enable us to describe the structure of atoms in the next chapter with greater understanding, and prepare us for later chapters.

Most radiographers will at one time or another have experienced the effects of so-called 'static' electricity generated by friction of one material on another. A common example occurs during the removal of items of clothing such as a nylon sweater. Pulling off the sweater may be associated with crackling noises and tingling sensations which may be quite unpleasant. Changing pillow cases is another example where frictional electricity may be produced. Even walking around on certain types of floor covering may cause an accumulation of electric charge on one's body which could later discharge in a disturbing way. Such occurrences are common in the dry atmosphere of hospitals. The discharge sparks may be large enough to be visible and may cause quite serious problems; e.g. in operating theatres where explosive anaesthetic gases may be present, and in x-ray darkrooms where discharges onto an x-ray film may cause annoying 'static marks' on the image.

To investigate these effects we need a rather more reliable method of generating frictional electricity. Some materials will allow electric charges to flow easily through them and are known as electrical conductors, e.g. iron, aluminium, copper. Other materials do not permit electric charges to flow through them. They are called electrical insulators, e.g. rubber, polythene, nylon, glass. Generally the effects of frictional electricity are associated with insulators rather than conductors of electricity. Combing one's hair with a nylon comb will invariably enable the comb to attract small pieces of paper. Even better, if we rub a polythene rod with a nylon headscarf the rod becomes capable of attracting the small pieces of paper. In other words, the rubbing action has produced some change in the rod giving it its property of attraction. We explain this by saying that the rod has become *charged* with electricity and it is the electric charge which is causing the attraction. It is also found that the nylon headscarf acquires the ability to attract things—the nylon, too, has become charged with electricity.

TYPES OF ELECTRIC CHARGE

Experiments confirm that there are two kinds of electric charge. They are called *positive* (the type on the polythene rod) and *negative* (the type on the nylon).

In its normal state all matter contains lots of electric charges, but for every positive charge there is a corresponding negative charge, and their effects exactly balance. Since there is no majority of either positive or negative charge we say the matter is *neutral*.

When some materials are rubbed together, the friction generated may cause some charges (particularly negative charges) to be transferred from one material to the other. In the case of the polythene rod and the nylon headscarf, negative charges are transferred from the rod to the headscarf. The state of neutrality is then disturbed, and the rod will have a majority of positive charges while the nylon will possess a majority of negative charges (Fig. 3/1). It is usually the negative charges which are transferred because the positive charges are too firmly fixed to move.

ELECTRIC FORCE

We have explained that electrically charged objects are able to attract small pieces of paper. This implies that forces must be

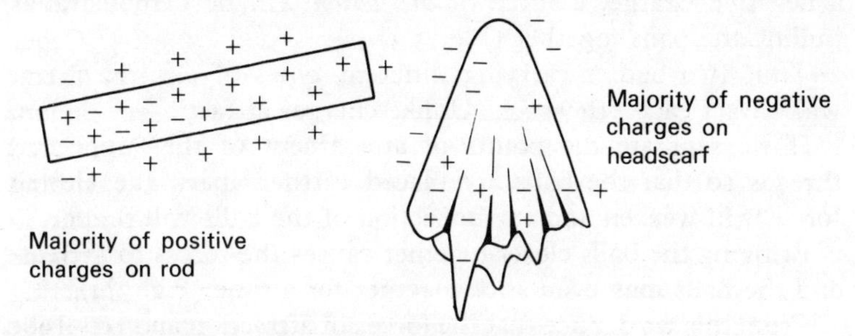

Fig. 3/1. Charged polythene rod and nylon headscarf.

involved, forces which are not present when the objects are in a neutral state. Experiments can be carried out to investigate the properties of these electric forces. Because the forces are very weak, only light objects will be affected enough for the forces to be demonstrated adequately.

Consider, therefore, two small balls made of crumpled aluminium foil ('silver paper' as found wrapped around blocks of chocolate is ideal) each about 1 centimetre in diameter, suspended on fine thread (nylon fishing line). If both balls are given a positive charge by touching them with a charged polythene rod a force of *repulsion* will be seen to exist between them, pushing the balls apart. Charging both balls with a negative charge would produce the same effect (Fig. 3/2).

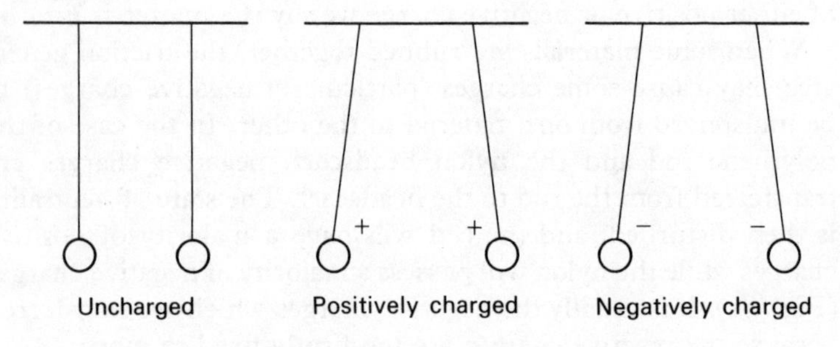

Fig. 3/2. Repulsion between like charges.

In other words, two bodies carrying the *same* type of electric charge will repel each other: i.e. 'Like charges repel'.

However, if one ball is given a positive charge and the other a negative charge, a force of *attraction* will be demonstrated, pulling the balls together (Fig. 3/3).

Thus two bodies carrying different types of electric charge will attract each other: i.e. 'Unlike charges attract'.

If we separate the points of attachment of the supporting threads so that the balls are placed further apart, the electric forces will weaken and the deflection of the balls will reduce.

Bringing the balls closer together causes the forces to increase and the balls may even stick together for a time (Fig. 3/4).

From this we deduce that the forces of attraction and repulsion are related to the *separation distance* between the charged objects.

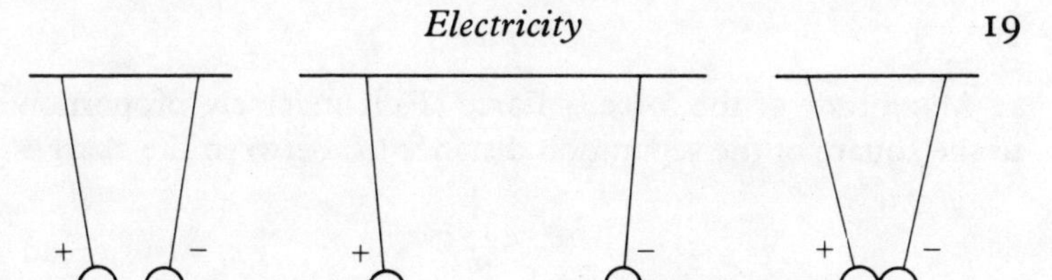

Fig. 3/3. Attraction between unlike charges.

Fig. 3/4. The effect of distance on electric force.

In fact, it is found that the force (F) between two charges is inversely proportional to the square of the distance (d) between them; i.e.

$$F \propto \frac{1}{d^2}.$$

This is an example of the relationship known as the Inverse Square Law which we shall meet again in Chapters 8 and 14. So, if we double the separation, the force would be four (2^2) times weaker. If we treble the separation, the force would be nine (3^2) times weaker, and so on.

As we have already explained, electric forces are generally very weak, but they may become strong enough to be significant when the separation distances are very small.

We can also influence the strength of the electric forces between charged objects by varying the *quantity* of charge involved. If the aluminium foil balls are given a large quantity of charge, they will attract or repel each other more strongly than if only a small amount of charge had been involved. If the quantities of charge on the balls are q_1 and q_2, the force (F) between them is proportional to the product of the charges; i.e. $F \propto q_1 q_2$.

Let us now summarise what we have learnt about the electric forces between charges.

SUMMARY OF THE LAWS OF ELECTRIC FORCE

1. Direction of the force. A force of repulsion operates between like charges. A force of attraction operates between unlike charges.

2. Magnitude of the force. Force (F) is inversely proportional to the square of the separation distance (d) between the charges:

$$F \propto \frac{1}{d^2};$$

and force (F) is directly proportional to the product of the quantities of charge (q_1 and q_2):

$$F \propto q_1 q_2 .$$

These two relationships can be combined into one:

$$F \propto \frac{q_1 q_2}{d^2}.$$

There is one other factor which has an influence on the strength of the electric forces. It relates to the *medium* in which the forces are acting; i.e. in our experiment with aluminium balls it relates to the substance (air) between the balls. If it were practical to replace the air with some other medium, e.g. oil, we would find that the magnitude of the electric forces would be different. We call this property of a medium which influences electric forces, *permittivity* or the *dielectric constant* of the medium.

ELECTRIC FIELDS

To help us to explain the behaviour of electric charges we invent the idea of an electric 'force field' surrounding electric charges. We can define an electric field as being an area or region in which an electric charge will experience an electric force.

To indicate the direction of the electric force, we use the idea of a *line of force*. This is defined as the path a positive charge would follow if it were free to move. Fig. 3/5 illustrates the lines of force around a fixed positive electric charge. They point *away* from the fixed charge because a free positive charge would be *repelled* away in these directions. The lines of electric force around a fixed negative charge would point *towards* the charge because a free positive charge would be *attracted* towards it.

We can also indicate the magnitude of the electric force (called the 'field strength') by the concentration of lines of force. If a high concentration of lines of force exist, the electric fields are strong. If a low concentration of lines is shown, the forces are weak. Referring again to Fig. 3/5, we can see that the concentration

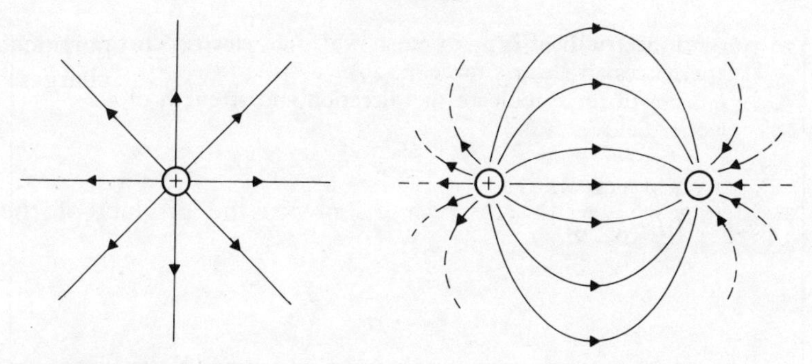

Fig. 3/5. Lines of force round a positive charge.

Fig. 3/6. Electric field between two unlike charges.

of lines of force is greatest close to the fixed charge, and reduces as we move further away. This supports what we have already seen of the effect of distance on the magnitude of electric forces. Thus the pattern of the lines of electric force around a charge map out the electric field.

The electric field produced by more than one charge is rather more complicated, but it is still defined in the same way, by deducing what a free positive charge would do if placed in the electric field. An example is given in Fig. 3/6. This shows that a free positive charge would always be drawn towards the negative charge—it is guided there by the electric field. If a free negative charge were put in the field, it would similarly be guided but to the positive charge. This use of an electric field to guide or focus charges is used in the x-ray tube and will be discussed in Chapter 12.

CHAPTER SUMMARY

1. Friction causes static electricity to be generated (p. 16).
2. There are two types of electric charge, called positive and negative (p. 17).
3. Neutral objects contain equal numbers of positive and negative charges (p. 17).
4. Like charges repel each other (p. 18).
5. Unlike charges attract each other (p. 18).

6. An electric field is a region in which an electric charge experiences an electric force (p. 20).
7. Lines of force indicate the direction and strength of an electric field (p. 20).

KEY RELATIONSHIP

$F \propto \frac{q_1 q_2}{d^2}$ (p. 20).

Chapter 4
Atomic structure

Later in our studies, in Chapters 16 and 17, we shall examine what happens when a beam of x-rays is directed at matter, particularly living tissue. In order to explain the processes which take place we need to have a basic knowledge of the way in which matter is constructed. This knowledge will also help us to understand how electricity flows through matter, how x-rays are produced in the x-ray tube and how gamma rays are produced by radioactive materials.

ELEMENTS AND COMPOUNDS

All matter is made up of chemical substances of two basic kinds:

1. Elements. These are chemicals that cannot be broken down into simpler chemical forms; e.g. hydrogen, carbon, oxygen, nitrogen.
2. Compounds. These are the result of two or more elements linking together chemically; e.g. water is a compound of the elements hydrogen and oxygen; the anaesthetic gas nitrous oxide is a compound of the elements nitrogen and oxygen.

We may also find matter composed of a mixture of elements or compounds; e.g. air is a mixture of the elements nitrogen and oxygen; i.e. the oxygen and nitrogen are not linked together at all.

ATOMS AND MOLECULES

From what we have learnt so far it follows that an element is the simplest form in which matter exists. If we have a sample of an element e.g. a strip of lead, and cut it into two, the resulting pieces will still be made of lead, the same element. If we go on doing this repeatedly so that our samples get smaller and smaller we will eventually reach the stage where it would be impossible to cut the samples into smaller pieces. We would have reached the smallest particle of the element which can exist and still retain all the features of the element. This is called the *atom*. The original strip of lead which we used as an example is made up of millions upon millions of atoms. In fact, all matter is composed of atoms.

ATOMS

An atom is defined as the smallest part of an element which retains the chemical properties of the element. Atoms are incredibly small—about one ten-millionth of a millimetre across (10^{-10} m).

If we cut up a sample of a compound rather than an element we will eventually reach a stage where to cut any further would separate the compound into the individual elements of which it is made. The smallest part of a compound which retains its original properties is called a *molecule*.

MOLECULES

A molecule is defined as the smallest part of a compound which retains the chemical properties of the compound. A molecule is made up of a number of atoms linked together; e.g. a water molecule is formed when two hydrogen (H) atoms and one oxygen (O) atom combine together (H_2O). This is a relatively simple molecule, but some molecules important in living organisms are highly complex and may contain many hundreds of atoms. It is also possible for atoms of the same element to combine together to form a molecule of an element; e.g. nitrogen (N) gas normally exists in the form of molecules, each containing two nitrogen atoms combined together (N_2).

To understand the behaviour of matter we need to examine the way in which atoms themselves are made.

THE STRUCTURE OF THE ATOM

Early this century a Danish scientist, Niels Bohr, developed some ideas on how an atom is constructed. We shall use Bohr's theory as the basis of our explanation. Bohr thought of the atom as being essentially electrical in nature. He visualised an atom consisting of minute particles—the so-called fundamental or elementary particles—held together by electric forces.

At the centre of an atom is the *nucleus*, about one ten thousandth of the diameter of the atom. Circulating round the nucleus at varying distances, like planets round the sun, are even smaller particles, the *electrons*. The nucleus carries a positive electric charge and the electrons possess a negative electric charge. The electric force of attraction between the nucleus and the electrons prevents the smaller particles from flying off. The roughly circular paths taken by the electrons around the nucleus are called orbits or shells, representing 'layers' at different distances from the nucleus. Fig. 4/1 shows this arrangement but we must remember that an atom is a three-dimensional structure which we have tried to represent in two dimensions on our diagram.

The nucleus is not a single particle, but consists of two types of particle packed together, and they are:

1. Protons—which each carry a *positive* charge equal and opposite to the charge on an electron.
2. Neutrons—which carry *no* electric charges (neutral).

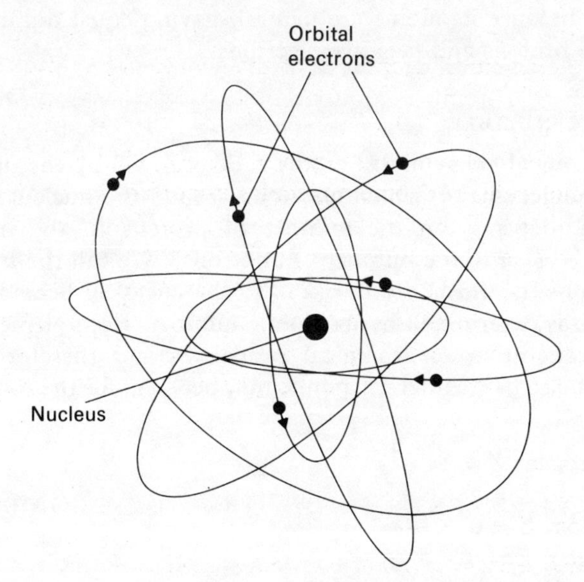

Fig. 4/1. An atom of carbon.

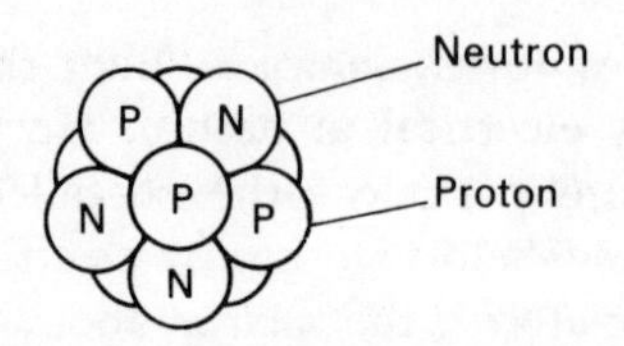

Fig. 4/2. The nucleus of an atom.

If positive protons are packed together, the electric forces between them will be very strong and will tend to cause the protons to fly apart ('like charges repel'). Neutrons appear to prevent this from happening and 'glue' the protons together (Fig. 4/2).

To summarise, three types of particle are found in atoms:

(a) Protons;
(b) Neutrons;
(c) Electrons.

These are the *fundamental particles.*

Most of the mass of an atom is concentrated in its nucleus. Protons and neutrons have about the same mass whereas an electron is nearly 2000 times lighter (proton mass = 1.66×10^{-27} kg).

As we said in Chapter 3, matter is generally electrically neutral. This is because its atoms are neutral, having equal numbers of positive protons and negative electrons.

ATOMIC NUMBER

The chemical behaviour of atoms is determined by the number and arrangement of orbital electrons around the nucleus. This, in turn, depends on the number of protons in the nucleus (which is *equal* to the number of electrons). We call the number of protons the *atomic number* (Z). So the chemical behaviour of an atom is determined by its atomic number. Each element has a different group of chemical properties and therefore each element also has a different atomic number; e.g. for the following elements:

Hydrogen, $Z = 1$
Helium, $Z = 2$
Carbon, $Z = 6$
Oxygen, $Z = 8$
Lead, $Z = 82$

Elements which are known have atomic numbers ranging from 1 to over a hundred. However, none of the elements with atomic numbers above 92 occur naturally but have to be manufactured artificially; e.g. plutonium ($Z = 94$) is manufactured in a nuclear reactor. Elements can be set out in a list, in the order of their atomic numbers, forming what is called the *Periodic Table* of elements.

MASS NUMBER

Although the number of protons in the nucleus of an atom determines its chemical properties the number of neutrons in an atom does not. It is possible to vary the number of neutrons without in any way disturbing the chemical behaviour of the atom. However, the *mass* of the atom *would* be affected. We call the total number of particles in the nucleus (i.e. the number of protons plus the number of neutrons) the *mass number* or *atomic mass* of the atom. The following are examples of elements with their individual mass numbers:

Hydrogen (H), mass number	= 1	(there are normally no neutrons in the hydrogen nucleus);
Helium (He), mass number	= 4	(2 protons + 2 neutrons);
Carbon (C), mass number	= 12	(6 protons + 6 neutrons);
Oxygen (O), mass number	= 16	(8 protons + 8 neutrons);
Lead (Pb), mass number	= 207	(82 protons + 125 neutrons).

ISOTOPES

Although all atoms of one element must have the same atomic number, they do not necessarily all have the same mass number; e.g. some atoms of hydrogen have a mass number of two instead of one. (These atoms of hydrogen are heavier than the usual kind because they contain a neutron in the nucleus and are called 'heavy hydrogen'. However, they combine with other chemicals in exactly the same way as 'normal' hydrogen). Atoms such as these, having the same atomic numbers but different mass numbers are called *isotopes* of an element.

There is a useful 'shorthand' method of writing down the numbers of particles in the nucleus of an atom, using the chemical symbol and two prefixes; e.g. $^{1}_{1}H$ represents normal hydrogen; $^{2}_{1}H$ represents heavy hydrogen. The upper prefix indicates the mass number, while the lower prefix shows the atomic number. The letter is the chemical symbol for the element. Other examples are: $^{4}_{2}He$, $^{16}_{8}O$, $^{12}_{6}C$, $^{207}_{82}Pb$. Specific arrangements of nuclei such as these are called *nuclides*.

Sometimes the electric forces of repulsion inside a nucleus are not completely overcome by the neutrons. Sooner or later such nuclides break up. They may eject particles and/or radiation energy from their nuclei. We call this process of nuclear disintegration *radioactivity* and the unstable nuclides are known as *radionuclides, radioisotopes* or *radioactive isotopes*. Wc shall be studying radioactivity in more detail in Chapter 20.

ELECTRON SHELLS

So far, we have been concentrating on the features of the nucleus of an atom but, as we have said, it is the number and arrangement of electrons orbiting around the nucleus which determines how an atom will form links with other atoms.

Electrons move in shells around the nucleus, at particular distances from it. The different shells represent different levels of potential energy, just as different heights above the ground represent different levels of gravitational potential energy. The further above the ground we go, the greater is the potential energy we are storing. In a similar way, the further an electron shell is from the nucleus, the greater is the potential energy of the electrons in that shell. However, in an atom, only certain specified energy levels are allowed; electrons cannot exist between these levels. The situation is similar to that of cars in a multi-storey car park; cars cannot be parked between floors, and electrons cannot be 'parked' between shells.

The energy levels (shells) are identified by letters: the K shell is the level closest to the nucleus, the next furthest is the L shell, then M, N, and so on. Again, as in a car park, each shell has a limited capacity for electrons, and the inner shells are always occupied first; e.g.

(a) the K shell can accept up to 2 electrons;
(b) the M shell can accept up to 8 electrons;
(c) the N shell can accept up to 18 electrons, etc.

Table 4/1 shows how the electrons are arranged in the first twelve elements in the Periodic Table.

Table 4/1. Electron arrangement of atoms.

Element	Atomic No.	No. electrons in shell			
		K	L	M	N
Hydrogen	1	1			
Helium	2	2			
Lithium	3	2	1		
Beryllium	4	2	2		
Boron	5	2	3		
Carbon	6	2	4		
Nitrogen	7	2	5		
Oxygen	8	2	6		
Fluorine	9	2	7		
Neon	10	2	8		
Sodium	11	2	8	1	
Magnesium	12	2	8	2	

Atoms 'prefer' to have shells either completely full or completely empty. They tend to become 'excited' if this state is not achieved; e.g. an element such as fluorine (Table 4/1) is just one electron short of completing its L shell. Atoms of fluorine, therefore, are always 'looking out for' an extra electron; i.e. fluorine is very reactive chemically.

On the other hand, an element such as lithium (Table 4/1) would be much happier if it were able to lose the sole electron in its L shell. Atoms of lithium are forever trying to give an electron away; i.e. lithium, too, is very reactive. Clearly if an atom of fluorine links up with an atom of lithium both atoms would be well satisfied. The result would be a molecule of the chemical known as lithium fluoride which we shall be meeting again in Chapter 19.

This example illustrates just one of the ways in which elements combine to form compounds: by sharing electrons. We can also see from this why it is mainly the arrangement of electrons in the *outer* shells of atoms that governs the basic chemical behaviour of elements. We can therefore group together those elements which have similar outer shells, and find that they have similar chemical properties. For example:

1. All elements like fluorine, whose atoms require just one electron to complete the outer shell, behave in a similar energetic way—known as 'halogens', they are fluorine, chlorine, bromine, iodine. They form a group of elements important in the chemistry of photography.
2. All elements with complete outer shells have no incentive to combine with other elements. These are all gases; classified as the 'inert gases', they are helium, neon, argon, krypton, xenon, radon.

IONISATION

As we pointed out at the beginning of this chapter, atoms are normally electrically neutral because the number of positive proton charges is exactly balanced by the number of negative electron charges in the shells around the nucleus. However, it may often be possible to remove an electron from the outer shell of an atom. If an electron is removed, the atom will no longer be neutral because there will be one positive charge in the nucleus which is not balanced by a corresponding negative charge. The atom is thus positively charged. We call a positively charged atom a positive *ion*.

The process of removing an electron from an atom is called *ionisation*. It is also possible to *add* an electron to a neutral atom, producing a negative ion.

BINDING ENERGY

To remove an electron we would have to overcome the electrical forces of attraction to the nucleus which keep the electrons in their shells. The quantity of energy required to remove an electron is called the binding energy of the electron. Each shell in a particular atom will have a different value of binding energy. The outer electron shells have weaker attraction forces acting than the inner shells because they are further away from the positive nucleus. Also the binding energy will depend on the quantity of positive charge in the nucleus, i.e. on the number of protons or atomic number. The higher the atomic number, the greater the binding energy of a particular shell.

To summarise, the binding energy of an electron depends on (a) which shell it is in; and (b) the atomic number of the atom.

IONISING RADIATION

Energy can be given to electrons to release them from their atoms by exposing the atoms to certain types of radiation. Such radiation is called ionising radiation, and important examples are x- and gamma rays.

With some materials it is possible to transfer electrons temporarily to higher energy levels (the atoms in effect store additional energy) in which case we say the atoms are excited. Eventually the atoms will be unable to 'contain themselves' any longer and will release the excess energy in the form of radiation (often visible light), as the electrons fall back to their original energy levels. An example of this process is seen in the use of fluorescent intensifying screens in diagnostic radiography. The x-radiation excites the atoms in the screens, and they then emit light which is used to expose the x-ray film and produce the radiographic image.

CHAPTER SUMMARY

1. All matter is made up of elements and compounds (p. 23).
2. An atom is the smallest part of an element (p. 24).
3. A molecule is the smallest part of a compound (p. 24).
4. The nucleus of an atom contains positive protons and neutral neutrons (p. 25).
5. Electrons orbit around the nucleus in shells (p. 25).
6. The atomic number of an atom is the number of protons in its nucleus (p. 26).
7. The mass number of an atom is the number of protons plus neutrons in its nucleus (p. 27).
8. Atoms having the same atomic numbers but different mass numbers are called isotopes (p. 27).
9. Radioactive isotopes have unstable nuclei (p. 28).
10. The chemical behaviour of atoms is determined by the electron arrangement of their outer shells (p. 29).

11. Elements can combine by sharing electrons (p. 30).
12. The process of removing an electron from an atom or adding an electron to an atom is called ionisation (p. 30).
13. The binding energy of an electron shell is the energy needed to release an electron (p. 30).

Chapter 5
Electric charge and potential

As we have seen in Chapter 4, the protons and electrons in an atom carry minute electric charges: electrons negative and protons positive. When we speak of a body being charged with electricity we mean it has a surplus or a deficiency of electrons so that its atomic charges no longer balance. A surplus of electrons would produce a net negative charge and a deficiency of electrons would give a net positive charge.

A possible (and logical) way of measuring electric charges would be to specify the *number* of electrons in surplus or deficit; i.e. to specify charge in terms of electronic charge units. In practice this is not done because the electron charge is so very small. It would be rather like measuring the distance from London to Paris in inches (about 14 million), rather than in miles (about 220). We therefore require a larger unit in which to measure quantities of electric charge. The SI unit of electric charge is called the *coulomb*, (C).

UNIT OF CHARGE

The coulomb is a unit derived from considering the force of attraction or repulsion between electric charges at a known distance apart. One coulomb is equivalent to the charge on about 6×10^{18} electrons. Even this represents quite a small quantity of charge in practical terms; e.g. an electric fire 'uses' 4 coulombs every second.

PROPERTIES OF ELECTRIC CHARGE

In Chapter 3, we have already seen the following:

1. Like charges repel and unlike charges attract.
2. Force between charges is proportional to $1/d^2$.
3. Force between charges is proportional to $q_1 \times q_2$.

We have also seen that charges can move or flow through conductors easily, and through insulators only with great difficulty. It follows that:

4. Charges (e.g. electrons) applied to a conductor will spread over its surface. They will not remain inside the conductor because of the mutual repulsion between the like charges (Fig. 5/1).
5. A charge put on to an insulator will remain where it is applied.

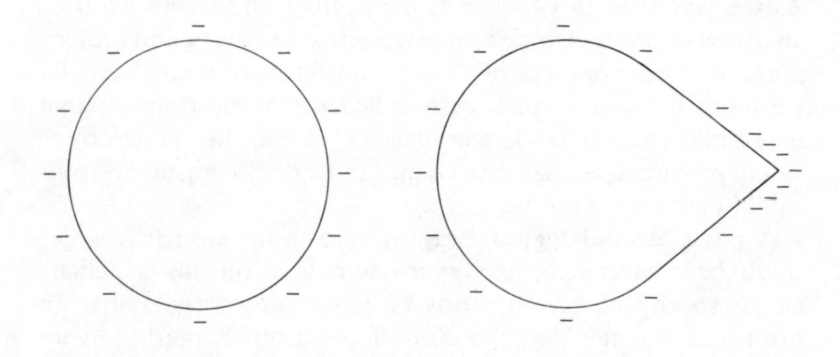

Fig. 5/1. Charge on the outside of a conductor.

Fig. 5/2. Charge concentrated on a *pointed* surface.

The charge on a pointed conductor tends to concentrate around the point, giving a stronger electric field the more curved or pointed the surface (Fig. 5/2). The field may become strong enough to *ionise* the air surrounding the point if electrons from the atoms of air are repelled away, and electrons from the charged conductor replace them. In this case the electric charge on the conductor will gradually leak away and we say an electric discharge occurs. Sometimes the discharge may be rapid and violent causing a spark to be produced. If the conductor is smoothly rounded rather than having sharp points or edges, discharge is not so likely to occur. When designing x-ray tubes, engineers make sure that parts which will be charged with high voltage electricity are as smooth as possible. The risk of electrical breakdown due to sparking is then reduced.

ELECTRICAL POTENTIAL ENERGY

As we have seen in Chapter 3, an electric charge which is in an electric field experiences a force (that is how we defined an electric field). In Fig. 5/3, if we wish to move the charge q to the left we will need to overcome the repulsive force due to the electric field, i.e. we will have to do *work*. If we succeed in moving the charge we shall have given it potential energy. If we then let it go it will move back, converting its potential energy into kinetic energy by virtue of its speed.

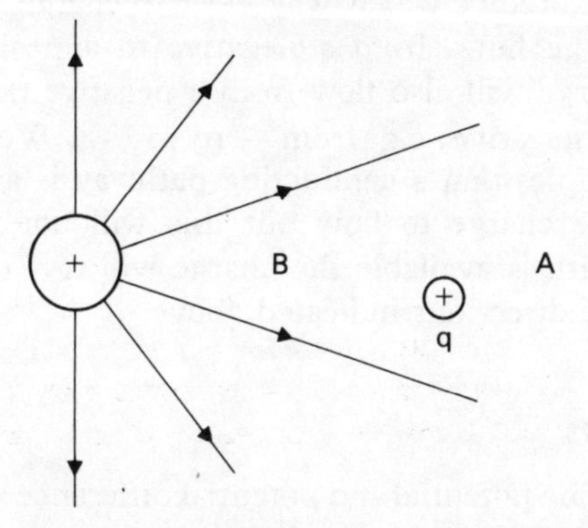

Fig. 5/3. Charge q in an electric field.

Electrical potential energy (usually called 'electrical potential' or just 'potential'), is thus very similar to gravitational potential energy: moving a charge *against* an electric field gives the charge stored energy; a charge moving *with* an electric field can convert its stored energy into useful work.

Electrical potential is always related to some arbitrary baseline, e.g. A in Fig. 5/3. The energy required (work done) to move the charge q from position A to position B is known as the potential difference (PD) between A and B. It can take on a positive or negative value according to whether energy has to be put in or energy is released. A useful baseline for potential is found by considering the charge q to start off at *infinity*, by which we mean some position far enough away for the force on the charge to be zero. If we suppose that A in Fig. 5/3 is at infinity, then it follows that the potential at A must be zero and a charge positioned there would have no stored energy. The work done in moving a charge from A to B would then be called the *potential* at B.

SUMMARY

To summarise our ideas on potential we could say that electrical potential, potential difference, or 'voltage' is an indication of the stored energy available. In general a high voltage electrical supply has a greater capacity for doing work than a low voltage supply.

It is also worth noting that a knowledge of potential difference enables us to predict which way electric charge will flow; e.g. negative charge flows from a negative to a positive potential. Negative charge will also flow from a negative potential to one which is less negative; e.g. from −10 to −4. We are assuming in these examples that a conducting pathway is available which will allow the charge to flow but this will not always be the case: if no path is available the charge will try (unsuccessfully) to flow in the directions indicated above.

DEFINITIONS

Before we define potential and potential difference we should note that because the energy required to move a charge will be greater for large values of charge than for small values, for the purpose of definition we assume that a 'unit charge' (i.e. 1 coulomb) is being considered.

1. The *potential* at a point is the work done in moving a unit positive charge from infinity to that point.
2. The *potential difference* between two points is the work done in moving a unit positive charge from one point to the other. (If the potential of two points is the same, the potential difference between them is zero).

UNIT OF POTENTIAL (AND POTENTIAL DIFFERENCE)

The SI unit of potential (and potential difference) is the volt, (V). The potential in volts at a point is the work done in joules in bringing one coulomb of positive charge from infinity to the point.

The potential difference in volts between two points is the work done in joules in moving one coulomb of positive charge from one point to the other.

There is thus a relationship between energy, electrical potential and charge:

$$\text{Energy} = \text{potential} \times \text{charge} \qquad (E = VQ)$$

$$\text{and} \quad \text{joules} = \text{volts} \times \text{coulombs}$$

THE ELECTRONVOLT

If we were to rearrange the definition of the volt we could arrive at a new definition of the unit of energy (the joule). For example we could say that one joule of work is done if a charge of one coulomb is moved through a potential difference of one volt. This is not an accepted definition since it defines the joule in terms of units which are derived from the joule. In other words we are using the joule to define the joule. However, it is a useful idea which is worth developing further.

Suppose that instead of moving one coulomb of charge, we move an electron charge (i.e. we move an electron) through a potential difference of one volt. Clearly the energy involved would be less than a joule because the charge on the electron is so much smaller than 1 coulomb. The quantity of energy thus defined is called an electronvolt (eV). It is a very small unit of energy:

$$1 \text{ eV} = 16.02 \times 10^{-20} \text{ J}.$$

Definition. An electronvolt is defined as the work done in moving an electron through a potential difference of 1 volt.

For example, if an electron 'falls' (is accelerated) through a PD of 1000 volts, it gains 1000 eV of kinetic energy, or 1 kiloelectronvolt (keV). A larger unit is the megaelectronvolt (MeV) where

$$1 \text{ MeV} = 1 \text{ million eV}.$$

The electronvolt is a very useful unit when working with vacuum tubes in which electrons are accelerated (e.g. an x-ray tube) and is generally used to describe the energy of the radiation produced in such devices.

EARTH POTENTIAL

If we add electrons to a body it becomes negatively charged. If we remove electrons it becomes positively charged. In the same sort of way if we add water to a tumbler which is half full, its level will rise; and if we remove water, its level will fall.

Suppose, however, we add a glass of water to the sea—does its level rise as a result? And if we remove a glass of water from the sea, does the sea level go down? Of course, the answer in each case is 'no' because the sea is so large that adding or subtracting a glass of water makes no significant difference to the sea level.

In the same way, adding or removing electric charges to or from the earth makes no significant difference to its potential. The earth's potential remains constant. We therefore use Earth Potential as a useful zero baseline from which to measure the potential of bodies and we can redefine potential as the potential difference of a body with respect to the earth.

EARTH CONNECTIONS

If we connect a body to the earth by an efficient conductor of electricity, the body will take on the same potential as the earth, (i.e. zero). This will make it electrically safe to handle. We call this 'earthing' or 'grounding' a component, and with electrical appliances which have metal outer casings, e.g. irons, vacuum cleaners, fridges, and x-ray sets, the parts which can be handled should *always* be earthed.

The connection to the earth is made in one of several ways:

1. Special earthing plates. These are metal conducting plates buried beneath the earth, having a large area of contact with moist soil. The electrical symbol for an earth connection is derived from this idea (Fig. 5/4).

2. Connections to water pipes. The mains cold water supply is usually provided through metal piping which enters a building from below ground. This can be used as an efficient earth connection.

What happens if an appliance is not earthed?

If it is working normally—nothing happens.

Fig. 5/4. Electrical symbol for an earth connection.

But, if a fault occurs, and a live (electrified) component comes into contact with the metal casing, the casing becomes 'live', i.e. its potential will not be zero. If we touch the casing *we* become its earth connection and electrons flow through *us* giving us an electric shock which may be mild or severe according to the rate of flow of electrons.

However, if the appliance had been properly earthed, the electrons would have run down to earth via the earth connections rather than through us because the human body is not a very efficient conductor of electricity. The usual result in practice is that a fuse blows, switching off the supply and making the appliance safe.

CHAPTER SUMMARY

1. The SI unit of electric charge is the coulomb (p. 33).
2. Charge concentrates round a pointed conductor (p. 34).
3. The potential at a point is the work done in bringing a unit positive charge from infinity to that point (p. 36).
4. The potential difference between two points is the work done in moving a unit positive charge from one point to the other (p. 36).
5. The SI unit of potential is the volt (p. 36).
6. The electronvolt is a unit of energy used in radiation physics (p. 37).
7. Earth potential is the baseline from which electrical potential is measured (p. 38).
8. Earthing an electrical appliance helps to make it safe to operate (p. 38).

KEY RELATIONSHIPS

Energy = potential × charge ($E = VQ$).
Volts = joules/coulombs.

Chapter 6
Conduction of electricity and storage of charges

In this chapter we shall extend our knowledge of the electrical properties of materials and examine some methods of using these materials to store electric charges.

ELECTRICAL CONDUCTORS AND INSULATORS

Early in Chapter 3 we defined electrical conductors as being materials through which electric charges move easily. Electrical insulators were said to be materials through which electric charges are not able to move.

Why do materials have these properties?—To explain this we need to remind ourselves of the electrical nature of matter that we studied in Chapter 4. Electric charges are either positive (carried by protons in the nuclei of atoms); or negative (carried by electrons). Electrons are much lighter than protons, and are less fixed in position so when electric charges are transferred from one place to another it is usually *negative* charges carried by electrons which move.

CONDUCTORS

A conducting material, therefore, must contain a lot of very loosely bound electrons, i.e. those which can easily be moved from one atom to another in the material. These electrons, which have very low binding energies, are able to drift from one atom to another quite freely. They are often referred to as 'free' electrons.

The direction of drift of an electron in a conducting material is entirely random. Just as many electrons drift to the left as to the right; just as many drift upwards as downwards and so on.

This is illustrated in Fig. 6/1. If, however, an electrical potential difference is applied across the conductor, all the electrons will tend to drift in the same direction—from the low potential to the high potential. Thus electric charge is flowing through the conductor and we say an electric current is flowing (Fig. 6/2). Many metals are good electrical conductors, silver, copper and aluminium being particularly efficient.

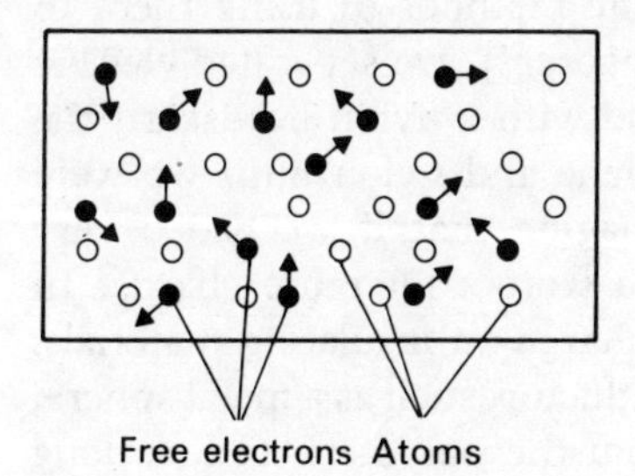

Fig. 6/1. Random drift of free electrons in a conductor.

Fig. 6/2. Free electrons in a conductor drifting towards a positive potential.

INSULATORS

Insulating materials, sometimes known as non-conductors or dielectric materials, do not allow electric charge to flow. This is because such materials possess very few 'free' electrons; the electrons are tightly bound and the application of a potential difference will not cause them to move. Glass, rubber and most plastics are good insulators. If a very large potential difference is applied even the tightly bound electrons may be drawn from their atoms and conduction may begin to occur. In these circumstances we say the insulation has broken down.

SEMICONDUCTORS

Some materials possess a limited number of free electrons and therefore will allow *some* flow of electric charge if a potential difference is applied. Such materials are known as semiconductors and they can be made to have some rather special and very useful electrical properties. Examples are the elements silicon and germanium. We shall be considering semiconducting materials in more detail in Chapter 13.

STORING ELECTRIC CHARGE

Now that we have an idea of the nature of conducting and insulating materials we can examine methods of using them to store electric charges. When, in Chapter 3, we separated electric charges by rubbing a polythene rod with a nylon headscarf, the charges were stored on the polythene and nylon until we were ready to use them. We might imagine that if we had a very large rod and a big scarf we could store a lot more charge. In this case we would be storing the charge on insulating materials. We could also store charge on a conductor, such as a metal sphere, by transferring charge on to it from the polythene rod, as long as the conductor was insulated from the ground so that the charges could not flow away to earth (Fig. 6/3).

As we transfer more and more charge onto the conductor its electrical potential rises until it becomes impossible to transfer any more charge on to the conductor because of the repulsion forces caused by the charges already on it. The potential of the

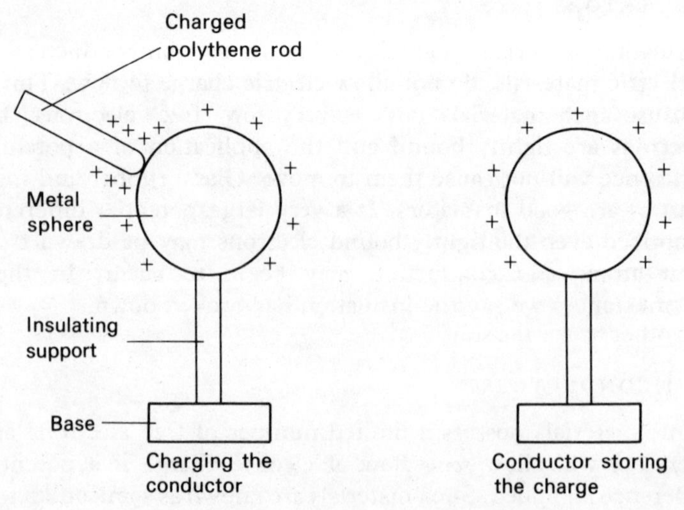

Fig. 6/3. Storing charge on a conductor.

conductor is then equal to the potential of the rod, and charge no longer flows from the rod to the conductor. The only way to store more charge would be to raise the potential of the rod in some way, e.g. by rubbing it more vigorously with the nylon. However, if the conducting sphere was bigger we could store more charge on it without having to increase the rod potential.

A useful analogy to consider is the storage of oxygen gas in the familiar black and white steel pressure cylinders which you will have seen about your hospital. Oxygen can be forced into the cylinder under pressure until the pressure of the gas inside the cylinder is equal to the pressure being used to put the gas in. The only way to store more oxygen is either to increase the pressure being applied, or to use a bigger cylinder.

CAPACITANCE

We could compare the storage capacities of different conductors by applying the same potential, say 1 volt, to each, and seeing how much charge in coulombs each conductor would accept. This is exactly how we define the *capacitance* (storage ability) of a conductor and we term the unit of capacitance the *farad* (F).

THE FARAD

The capacitance of a conductor in farads is defined as the quantity of electric charge, in coulombs, which it will store at a potential of 1 volt. It is also the charge which will *change* its potential by 1 volt.

For example, if a conductor is able to store 5 coulombs of charge at a potential of 1 volt, its capacitance must be 5 farads. The same conductor would be able to store 10 coulombs of charge at a potential of 2 volts; 15 coulombs at 3 volts, and so on.

In other words the stored charge (Q) is proportional to potential (V), or

$$Q \propto V.$$

Putting this in the form of an equation:

$$Q = \text{constant} \times V.$$

The constant of proportionality (C) is what we have termed capacitance; i.e.

$$Q = CV.$$

This is a very important relationship in considering the storage of electric charges.

The unit of capacitance which we have defined, the farad, proves in practice to be a very large unit and we find that the capacitances of conductors are measured in millionths of farads (microfarads), or even in millionths of microfarads (picofarads).

$$1 \text{ microfarad } (\mu F) = 10^{-6} \text{ F}$$

$$1 \text{ picofarad } (pF) = 10^{-12} \text{ F}.$$

CAPACITORS

We have described the storage of charge on a spherical conductor, but in practice such designs are very poor for storing charge; e.g. a 2-cm diameter metal sphere will only store 10^{-12} coulombs of charge at 1 volt. Its capacitance is only 1 pF. A much better arrangement is to have two conductors (electrodes) separated by an insulating material, one of the conductors being connected by a wire to earth. This arrangement is called a capacitor (or condenser) and is a device designed for the purpose of storing electric charges (Fig. 6/4). It is very much easier to transfer charges on to the left-hand conducting plate than it would be if the right-hand plate was absent or not earthed. This is because

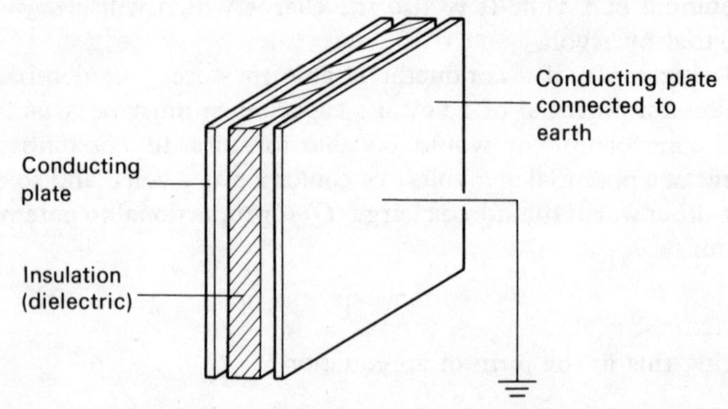

Fig. 6/4. Design of a simple capacitor.

the repulsion effect of charges already on the left-hand plate is weakened by the presence of opposite charges on the right-hand plate producing an attraction effect. As Fig. 6/5 shows, if the left-hand plate is being loaded with negative charges, the right-hand plate will become positively charged, and vice versa.

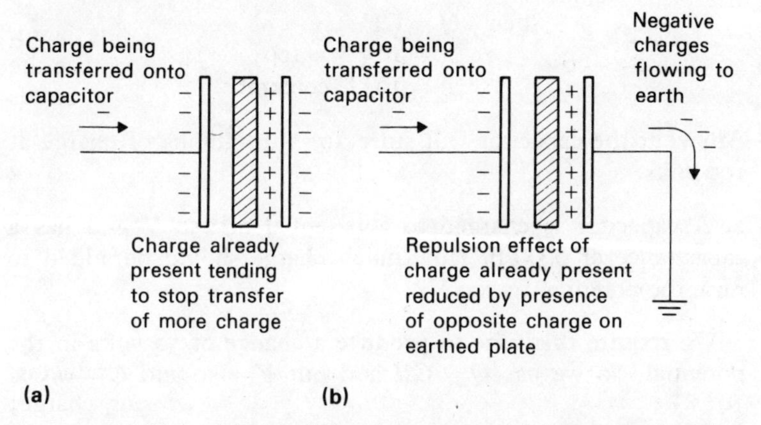

Fig. 6/5. (a) If the right-hand plate is unconnected, it has no effect since the attraction force of its positive charges is balanced by the repulsion force of its negative charges. The situation is therefore the same as that of a single conductor.

Fig. 6/5. (b) If the right-hand plate is earthed, its negative charges flow away to earth while its positive charges help to overcome the repulsion effect of the charges on the left-hand plate.

Because it is easier to charge the conductor in the way we have described, the work done in putting charges on it is less, so its potential is less. If charge can be put on with a smaller potential rise, then the capacitance of the device must be greater than that of a single conductor. In practice, the capacitance is up to a million times greater.

The relationship $Q = CV$ still applies, as it did to a single conductor, and Q represents the charge on *one* of the plates. It may be useful at this point to run through a few worked examples using this relationship.

WORKED EXAMPLES

1. A capacitor with a capacitance of 1 pF is charged to a potential of 100 V. How much charge will be stored at this potential?

Before simply inserting the values of capacitance (C) and potential (V) into $Q = CV$, we must stop and recall that Q, C and V should be in SI units, i.e. coulombs, farads and volts respectively. So we must first convert 1 pF into farads:

$$1 \text{ pF} = 10^{-12} \text{ F}$$

$$\text{then} \quad Q = CV$$

$$= 10^{-12} \times 100$$

$$= 10^{-10} \text{ coulombs.}$$

Answer: the capacitor will store 10^{-10} coulombs of charge at 100 volts.

2. A capacitor is charged to a potential of 250 V and has a capacitance of 0.5 μF. How much charge should be added to raise the potential to 300 V?

We require this time to produce a change of 50 volts in the potential. So we use $Q = CV$ and put $V = 50$ and $C = 0.5 \times 10^{-6}$ F;

$$\text{then} \quad Q = 0.5 \times 10^{-6} \times 50$$

$$= 25 \times 10^{-6} \text{ coulombs} \quad \text{(25 millionths of a coulomb)}$$

Answer: 25×10^{-6} coulombs of charge must be added to raise the potential from 250 to 300 volts.

3. What potential would be needed to make a 1 μF capacitor store 1 coulomb of charge?

This time we are told the values of C ($= 10^{-6}$ F) and Q ($= 1$ coulomb). We therefore need to rearrange the relationship $Q = CV$ to enable us to calculate V. We do this by dividing both sides of the equation by C:

$$\frac{Q}{C} = \frac{CV}{C}$$

$$\text{thus} \quad \frac{Q}{C} = V$$

$$\text{then} \quad V = \frac{1}{10^{-6}}$$

$$= 10^{6} \text{ volts.}$$

Answer: 1 million volts (10^6 volts) would be needed to make a 1 μF capacitor store 1 coulomb of charge! Clearly, no ordinary capacitor would survive that sort of treatment, but this example does show that only comparatively small quantities of electric charge can, in practice, be stored in capacitors.

FACTORS AFFECTING CAPACITANCE

Various design features of capacitors can be used to improve their storage capacity:

1. If the conducting plates are made larger in area the capacitance will be increased;
2. If the distance separating the conducting plates is decreased, the capacitance will be increased;
3. If a more efficient dielectric material is used between the plates, the capacitance will be increased. The term *dielectric constant* is used to denote the effect on capacitance of the insulating layer between the plates.

In the parallel-plate type of capacitor we have been describing we can be more precise in stating the effect of these design features and we can say *exactly* how they influence capacitance:

1. Capacitance (C) is directly proportional to plate area (A): $C \propto A$.
2. Capacitance is inversely proportional to plate separation (d): $C \propto \frac{1}{d}$.
3. Capacitance is directly proportional to dielectric constant (k): $C \propto k$.

Combining these factors: $C \propto \frac{Ak}{d}$.

TYPES OF CAPACITOR

Frequently, capacitance is increased by having more than one pair of conducting plates. A whole series of plates may be used, effectively increasing the total plate area (Fig. 6/6). Such devices are called multileaf capacitors and their capacitance is proportional to the number of plates employed.

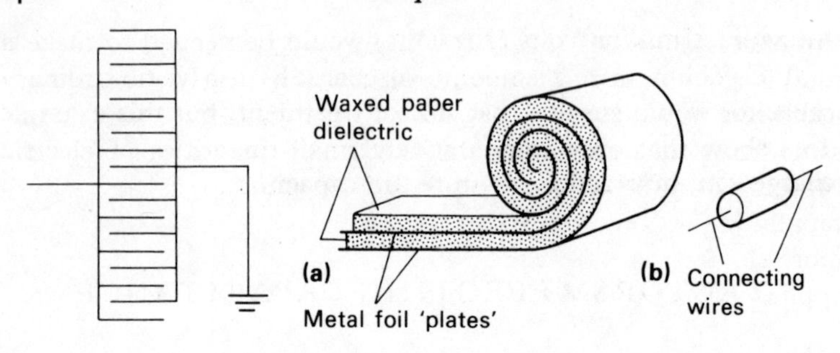

Fig. 6/6. Multileaf capacitor with eight pairs of plates.

Fig. 6/7. (a) The 'rolled sandwich' of a foil/wax paper capacitor. (b) The external appearance of the capacitor.

If a thin metal foil is used to form the conducting layers and thin, wax-impregnated paper to form the dielectric, the whole 'sandwich' may be rolled up into a 'Swiss roll' arrangement. In this way a large plate area can be combined with a small plate separation and made into a tiny electrical component which has a reasonably large capacitance (Fig. 6/7).

To produce very large values of capacitance, a device called an electrolytic capacitor may be used. Although it may have a large capacitance, e.g. 1000 μF, it can easily be damaged if it is wrongly connected in an electrical circuit.

CHARGING CAPACITORS

When a capacitor is being charged we find that the charge builds up rapidly at first, but after a while slows down, until eventually no more charge is being added. This is due to the changing potential on the capacitor. As we saw earlier in this unit in our analogy with the oxygen cylinder, we can only continue to 'charge' the cylinder as long as the pressure we are applying to put the oxygen in is greater than the pressure inside the cylinder. As the pressure inside builds up, we find it more and more difficult to force in any more oxygen.

When we charge a capacitor the difference in electrical potential is great at first, so charge builds up rapidly, but as time goes on the potential on the capacitor rises, the potential difference reduces, and the charging rate tails off to zero. We can illustrate this effect on a graph (Fig. 6/8). What we have termed 'charging rate' is the value of electric current flowing into the capacitor, i.e. the charging current. Fig. 6/9 shows another aspect of the same situation: it illustrates the rise in electrical potential on the capacitor as more and more charge is added. The potential rises rapidly at first (because charge is added quickly), but later rises more slowly and eventually reaches a steady value equal to the applied potential.

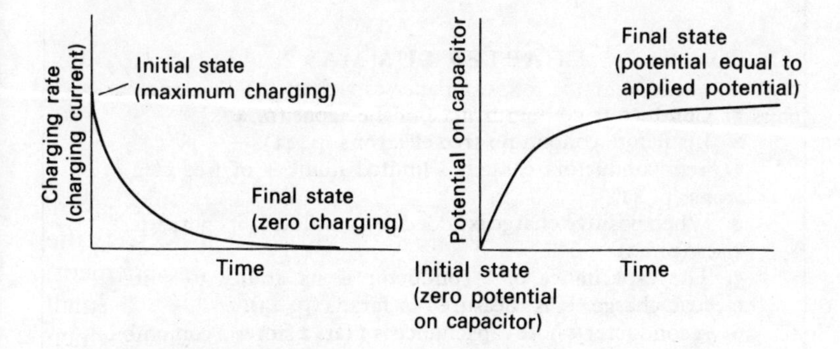

Fig. 6/8. Exponential decay of the charging rate of a capacitor.

Fig. 6/9. Exponential growth of the potential on a capacitor as it is charged.

The particular relationship between charging rate and time is called an exponential relationship because it can be described mathematically using a number called the exponential constant. The major feature of the relationship is that in equal intervals of time the variable (in this case charging rate) will decrease by equal fractions. This reducing relationship is known as exponential decay and we shall meet it again in different situations in Chapters 9, 16 and 20.

The relationship between potential and time illustrated in Fig. 6/9 is also an exponential one, but in this case the variable increases by equal fractions in equal intervals of time. This is known as exponential growth and we shall meet it again in Chapter 9.

The total time taken for a capacitor to reach a specified state of charge or a specified potential can be controlled by limiting the charging current. If the charging current is kept very small the capacitor will take a long time to charge, but if the restraint is removed it will charge quickly.

In Chapter 7 we shall consider some of the ways in which electrical components, including capacitors are connected in electric circuits and also the behaviour of capacitors when they discharge the electric charges they store.

CHAPTER SUMMARY

1. Conductors contain many free electrons (p. 40).
2. Insulators contain no free electrons (p. 41).
3. Semiconductors contain a limited number of free electrons (p. 41).
4. When positive charge is added to a conductor its potential rises (p. 42).
5. The capacitance of a conductor is its ability to store electric charge. It is measured in farads (p. 43).
6. A conductor whose capacitance is 1 farad stores 1 coulomb of charge at a potential of 1 volt (p. 43).
7. A capacitor consists of two electrodes separated by an insulator (p. 44).
8. Capacitors charge according to an exponential relationship (pp. 48–49).

KEY RELATIONSHIPS

$Q = CV$ (pp. 43–44).

$C \propto \frac{Ak}{d}$ (p. 47).

Chapter 7
Electric current

In Chapter 6 we discussed putting charge on capacitors; this *movement* of electric charge is known as *electric current.*

If we make one end of a conductor positive and the other end negative, electrons would move towards the positive end. It is this flow of electrons which constitutes the electric current.

The action of making one end of a conductor positive and the other negative is known as 'applying a potential difference'.

So we have established that electrons (electric current) flow from negative to positive, but when electric current was first observed physicists did not know about electrons, so they guessed (wrongly) that current flowed from positive to negative. Since this convention was used as the basis for some of the laws of physics we must differentiate between the two types of current, so we call flow from negative to positive *electron* current, and flow from positive to negative *conventional* current. In the rest of this book the word current should be taken to mean *electron* current, unless stated otherwise.

The unit of current is the 'ampere', often abbreviated to 'amp', or just 'A'. An 'amp', therefore, represents a certain charge, or number of electrons, flowing per unit time. If we used the number of electrons, the figures involved would be very large, so we use instead the *coulomb* (1 coulomb represents the charge on 6×10^{18} electrons.

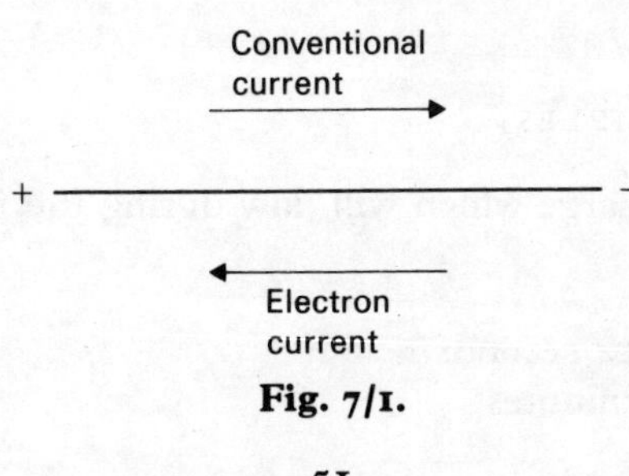

Fig. 7/1.

DEFINITION

The unit of current called the ampere is that current which flows when a charge of 1 coulomb passes a given point every second.

From this definition we can derive the following:

$$\text{Current} = \frac{\text{charge}}{\text{time}}.$$

If I = current in amperes; Q = charge in coulombs; and t = time in seconds; then

$$I = \frac{Q}{t},$$

or

$$1 \text{ amp} = \frac{1 \text{ coulomb}}{1 \text{ second}},$$

and

$$Q = It;$$

i.e.

$$\text{total charge} = \text{current} \times \text{time}$$

or

$$1 \text{ coulomb} = 1 \text{ amp} \times 1 \text{ second}.$$

We sometimes need to talk about currents smaller than 1 amp, so the term 'milliamp' is used, where:

$$1 \text{ milliamp} = 10^{-3} \text{ amp}.$$

or

$$1000 \text{ mA} = 1 \text{ amp}.$$

A device used to measure the flow of current is called an ammeter (or milliammeter).

WORKED EXAMPLES

Calculate the charge which will flow during the following x-ray exposures:

(1) 400 mA; 0.04 seconds
(2) 15 mA; 15 minutes

1. Using the equation $Q = It$ (remember Q is in coulombs, I is in amps, t is in seconds); therefore

$$Q = \frac{400}{1000} \times \frac{4}{100}$$

$$= \frac{1600}{100\,000}$$

$$= 0.016 \text{ coulomb}$$

$$= 16 \times 10^{-3} \text{ coulomb.}$$

2. Again using the equation $Q = It$;

$$Q = \frac{15}{1000} \times 15 \times 60$$

$$= \frac{13\,500}{1000}$$

$$= 13.5 \text{ coulombs.}$$

The charge flowing during an x-ray exposure is usually quoted in milliampere-seconds (mAs) where 1 mAs = $\frac{1}{1000}$ coulomb.

CIRCUIT SYMBOLS

When we draw electrical circuits (circuit diagrams) we use a selection of symbols, each of which represents a component, as

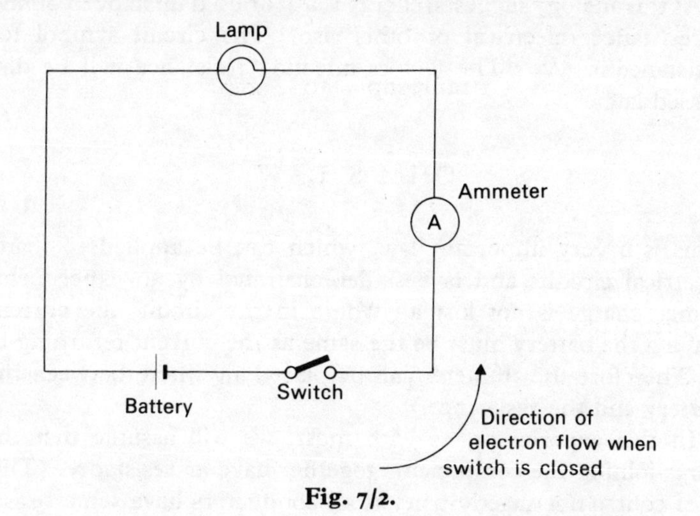

Fig. 7/2.

shown in Fig. 7/2. For a current to flow it is always necessary to have a *complete* circuit, i.e. a continuous path for the electrons to flow along. Thus the lamp in the diagram will only light up when the switch in the circuit is closed. A battery is used in the circuit as a source of potential difference, to drive the current through the circuit.

POTENTIAL DIFFERENCE

This can be thought of as the difference in *electrical pressure* between any two points in the circuit. As you would imagine, a 10-volt battery will supply twice the potential difference of a 5-volt battery. A device for measuring potential difference is called a voltmeter.

RESISTANCE

Resistance, as the word suggests, is the 'opposition' experienced by an electric current. For example, if you imagine walking down a wide corridor, there would be very little 'opposition' to your progress. Now imagine trying to walk down a crowded narrow corridor. This would be much harder, i.e. the narrow corridor would offer a higher resistance.

As this analogy suggests, energy is lost or used up in overcoming a resistance (electrical or otherwise). The circuit symbol for resistance is -ʌʌʌ-. The factors affecting resistance will be discussed later.

OHM'S LAW

This is a very important law, which can be applied to many electrical circuits and is best demonstrated by an experiment. (Since charge is not lost anywhere in the circuit, the current leaving the battery must be the same as the current returning to it. Therefore the ammeter can be placed anywhere between the battery and the resistance.)

In this experiment, as with most, we will assume that the wires joining the components together have *no* resistance. (This is of course not strictly true, as all conductors have some resist-

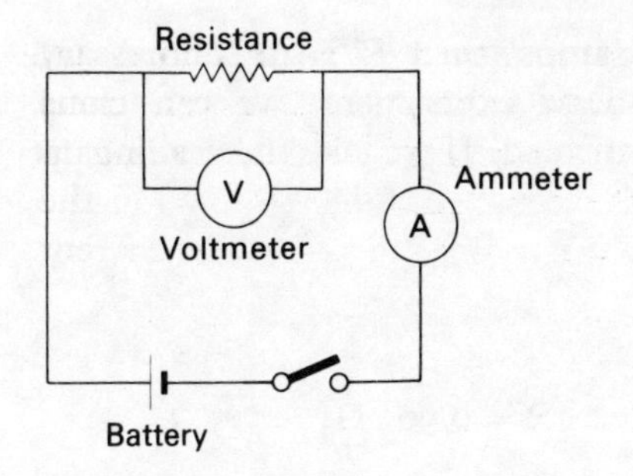

Fig. 7/3. Ohm's Law circuit.

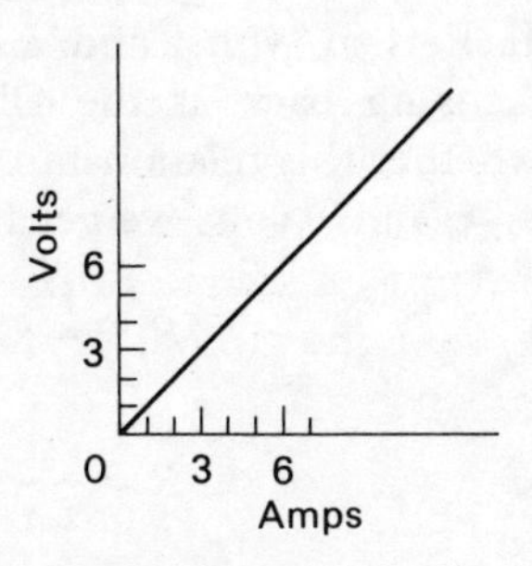

Fig. 7/4. Results of Ohm's Law experiment displayed as a graph of current against voltage.

ance.) By altering the values of voltage supplied by the battery the following values of current were obtained:

V	1	2	3	4	5	6	(volts)
I	1.1	2.2	3.3	4.4	5.5	6.6	(amps)

When these values are plotted in the form of a graph we get a straight line.

From similar results Ohm concluded that V was proportional to I and Ohm's Law is derived from this conclusion.

DEFINITION OF OHM'S LAW

Ohm's Law states that the current flowing through a conductor is directly proportional to the potential difference between its ends, so long as all physical conditions (e.g. temperature) remain constant.

i.e. $V \propto I$;

therefore $V = I \times \text{constant}$.

This constant is known as *resistance* (R) and is measured in 'ohms',

$$\text{i.e.} \quad V = I \times R,$$

$$\text{and} \quad R = \frac{V}{I}.$$

If V is in 'volts' and I is in 'amps' then R is in 'ohms' (Ω). Looking back at the Ohm's Law experiment, we can check whether this relationship is confirmed. If we take the readings at 1, 3, and 5 volts we get from

$$R = \frac{V}{I}:$$

$$R = \frac{1}{1.1} \quad \text{i.e.} \quad R = 0.909\ \Omega;$$

$$R = \frac{3}{3.3} \quad \text{i.e.} \quad R = 0.909\ \Omega;$$

$$R = \frac{5}{5.5} \quad \text{i.e.} \quad R = 0.909\ \Omega.$$

From $R = \frac{V}{I}$ we can define the 'ohm'.

DEFINITION OF THE OHM

If a potential difference of 1 volt drives a current of 1 amp through a conductor, the resistance of the conductor is said to be 1 ohm.

We often need much larger values, so we use 'kilohms' (kΩ) and 'megohms' (MΩ) where:

$$1\ \text{k}\Omega = 1000\ \text{ohms},$$

$$\text{and} \quad 1\ \text{M}\Omega = 1\,000\,000\ \text{ohms}.$$

WORKED EXAMPLES

1. If a potential difference of 6 volts is applied to a 3-kΩ resistance, calculate the current that will flow through it.

$$\text{From Ohm's Law:} \quad I = \frac{V}{R},$$

$$\text{therefore} \quad I = \frac{6}{3000}$$

$$= \frac{1}{500}\ \text{amps}$$

$$= 2\ \text{milliamps}.$$

2. If a current of 4 amps flows through a resistance of 60 Ω, calculate the potential difference across it.

From Ohm's Law: $V = IR$,

therefore $V = 4 \times 60$

$= 240$ volts.

FACTORS AFFECTING RESISTANCE

These can be discussed by comparing electricity with the flow of water through a pipe. For example:

1. Length. A long pipe will obstruct the flow of water more than a short one. This is analogous to a long conductor which has a greater resistance than a short one.

2. Cross-sectional area. A narrow pipe will obstruct the flow of water more than a wide one. This is analogous to a thin conductor which has a greater resistance than a thick one; i.e. if l = length, and A = cross-sectional area;

$$\text{then} \quad R \propto \frac{l}{A},$$

$$\text{or} \quad R = \frac{sl}{A} \quad \text{where } s \text{ is a constant.}$$

We call this constant 'resistivity'. It is a measure of the opposition a particular substance exhibits to the passage of electric current. It is defined as; the resistivity of a conducting material is its resistance in ohms, measured between opposite faces of a cube with a side length of 1 metre. For example the resistivity of copper is 1.72×10^{-6} ohm metres at 20°C. Thus, pieces of wire with the same length and cross-sectional area, if made of different materials, e.g. copper, silver, and gold, will have different values of resistance.

WHAT CAUSES RESISTANCE?

We mentioned earlier that resistance was the 'opposition to the flow of current'; but what produces this opposition?

If we imagine looking closely at a piece of wire we would see all the atoms vibrating and free electrons moving around at random. When we apply a potential difference across the wire the free electrons tend to move toward the positive end, but on the way they collide with some of the vibrating atoms; this effect is resistance.

WHAT HAPPENS IF WE HEAT THE WIRE?

When we apply heat to the wire the atoms absorb energy and vibrate more actively; thus it is more likely that the flowing electron will collide with the vibrating atoms, i.e. the resistance is *increased.*

Consider this analogy. Imagine you are in a room half-full of people and you are trying to walk from one side to another. If the people are moving slowly your path to the other side would be fairly easy, and this is comparable with low resistance. If, however, all the people were rushing around very actively (with lots of energy) it would be quite hard to reach the other side and this is comparable with high resistance.

SUMMARY

In the case of conductors, resistance *increases* with temperature.

CIRCUIT LAWS

These are rules which govern the operation of a circuit with several components.

COMBINATIONS OF POTENTIALS

If we connect two batteries together each supplying a PD of 1 volt, we find the total PD is 2 volts (Fig. 7/5). Similarly with four batteries (Fig. 7/6). This is called connecting batteries in *series.* The circuit symbol for a high voltage battery is ||- - - -||. If we connect two 1-volt batteries in *parallel,* i.e. joining positive to positive and negative to negative, (Fig. 7/7) we find we get a PD of 1 volt. But this arrangement can supply *twice* the current of a single battery.

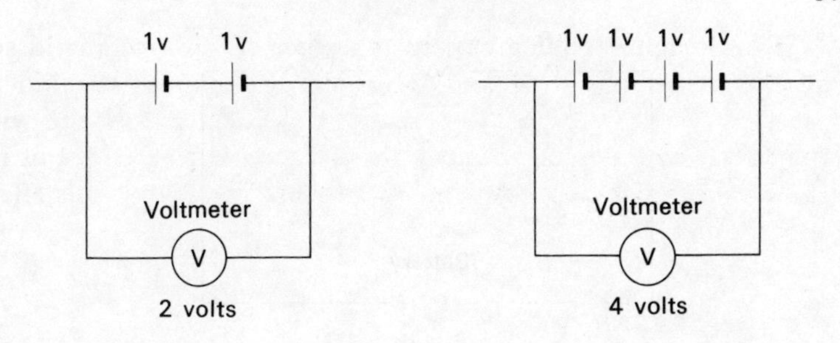

Fig. 7/5. Two batteries in series. **Fig. 7/6.** Four batteries in series.

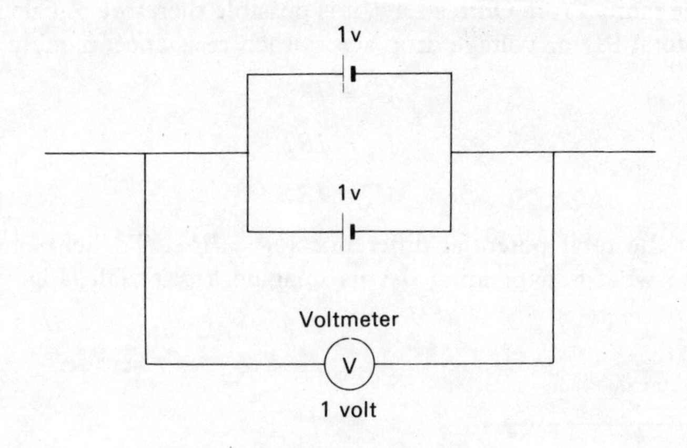

Fig. 7/7. Batteries in parallel.

COMBINATIONS OF RESISTANCES

Now let us look at a circuit with three resistances in series (Fig. 7/8). We have already said that resistance is the 'opposition to the flow of current'. It follows, then, that if we put three resistances in series the *total* value of these resistances can be calculated by *adding* their individual values. This is a very important idea and we will illustrate it with a numerical example in a moment.

It is first worth considering the potential at various points in this circuit. Since the current in the circuit is the same in all places, it follows that the current through each of the resistances

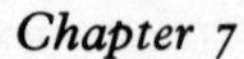

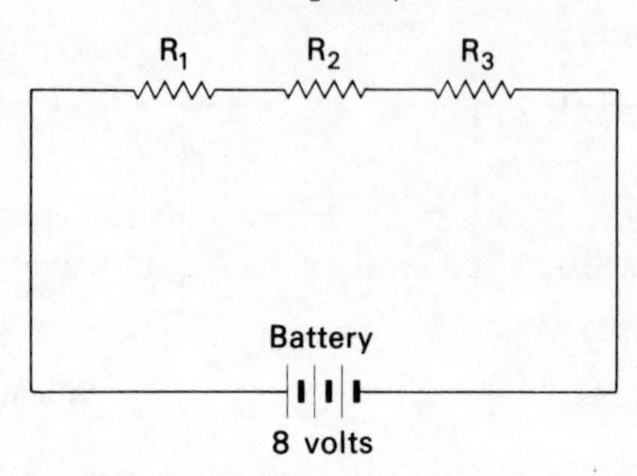

Fig. 7/8. Resistances in series (R_1, R_2 and R_3 are three resistances in series).

is the same. From Ohm's Law, it is possible therefore to calculate the total PD or voltage drop across each resistance, namely:

$$V_1 = IR_1$$

$$V_2 = IR_2$$

$$V_3 = IR_3$$

with the total potential difference $V_T = IR_T$. To help understand what is happening, let us imagine a waterfall (Fig. 7/9).

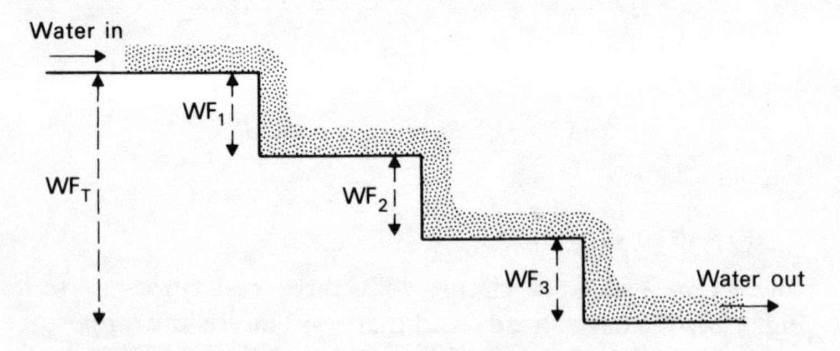

Fig. 7/9. Waterfall analogy of resistances in series.

The amount of water flowing in at the top is the same as the amount of water flowing out at the bottom ('current'). The total height difference is WF_T which is in turn made up of WF_1, WF_2 and WF_3. In this analogy we are comparing a drop in height with a drop in electrical potential.

WORKED EXAMPLE

If $R_1 = 3\ \Omega$, $R_2 = 5\ \Omega$, $R_3 = 8\ \Omega$, and the PD supplied by the battery is 8 volts; calculate (a) the total resistance (R_T); (b) the current (I); and (c) the PD across each resistance (V_1, V_2 and V_3).

(a) $R_T = R_1 + R_2 + R_3$

$R_T = 3 + 5 + 8$

$R_T = 16\ \Omega.$

(b) $I = \dfrac{V_T}{R_T}$ (where V_T is total PD)

$I = \dfrac{8}{16}$

$I = \dfrac{1}{2}$ amp.

(c) $V_1 = IR_1$

$V_1 = \dfrac{1}{2} \times 3$

$V_1 = 1\frac{1}{2}$ volts.

$V_2 = IR_2$

$V_2 = \dfrac{1}{2} \times 5$

$V_2 = 2\frac{1}{2}$ volts.

$V_3 = IR_3$

$V_3 = \dfrac{1}{2} \times 8$

$V_3 = 4$ volts.

If we add the individual potential differences we would expect to get the total potential difference across all three resistances (i.e. the PD supplied by the battery); i.e. $1\frac{1}{2} + 2\frac{1}{2} + 4 = 8$ volts.

RESISTANCES IN PARALLEL

As with the previous circuit, the current leaving the battery is the same as that returning to it, but in this case the current divides, some current passing through each resistance.

Since the potential difference across all the resistances is the same, the current divides up in a manner *inversely proportional* to their values. This is sensible when you think about it; because, given the choice of three routes, the majority of the current chooses the path of least resistance.

The total value of three resistances connected in parallel can be found by the following equation:

$$\frac{1}{R_T} = \frac{1}{R_1} + \frac{1}{R_2} + \frac{1}{R_3}.$$

So let us now calculate the total resistance of the three resistances shown in Fig. 7/10. If $R_1 = 2\ \Omega$, $R_2 = 3\ \Omega$, and $R_3 = 6\ \Omega$;

$$\frac{1}{R} = \frac{1}{R_1} + \frac{1}{R_2} + \frac{1}{R_3}$$

$$\frac{1}{R} = \frac{1}{2} + \frac{1}{3} + \frac{1}{6}$$

$$\frac{1}{R} = \frac{3 + 2 + 1}{6}$$

$$\frac{1}{R} = \frac{6}{6}$$

$$R = 1\ \Omega.$$

Therefore the current in the circuit will be

$$I = \frac{V}{R}$$

$$I = \frac{12}{1}$$

$$I = 12 \text{ amps.}$$

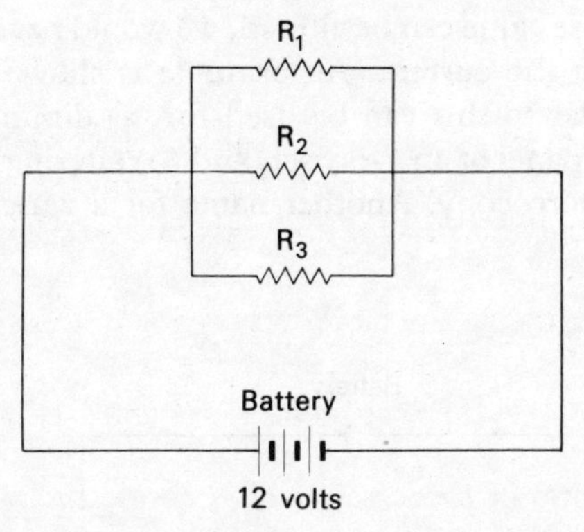

Fig. 7/10. Resistances in parallel (R_1, R_2 and R_3 are three resistances in parallel).

We can check our calculation by working out the current through each of the individual resistances:

$$I_1 = \frac{V}{R_1}, \quad I = \frac{12}{2} \qquad I = 6 \text{ amps.}$$

$$I_2 = \frac{V}{R_2}, \quad I = \frac{12}{3} \qquad I = 4 \text{ amps.}$$

$$I_3 = \frac{V}{R_3}, \quad I = \frac{12}{6} \qquad I = 2 \text{ amps.}$$

By adding these currents together, we get $6 + 4 + 2 = 12$ amps which equals the current we calculated using the total resistance value (see above).

SUMMARY

1. The current approaching resistances in parallel divides in inverse proportion to the value of the resistances.
2. The total value of resistances connected in parallel is always less than the value of any individual resistance.

USES OF A VARIABLE RESISTANCE IN AN X-RAY MACHINE

We have seen that the current flowing in a circuit is inversely proportional to the resistance $\left(I \propto \frac{1}{R}\right)$. If we put in a circuit a resistance whose value can be altered, we would have a convenient way of altering the current. An example is shown in Fig. 7/11. A circuit similar to this can be used for (a) dimming the lights on the control panel of an x-ray set, and (b) altering the screening mA during fluoroscopy. Another name for a variable resistance is a *rheostat*.

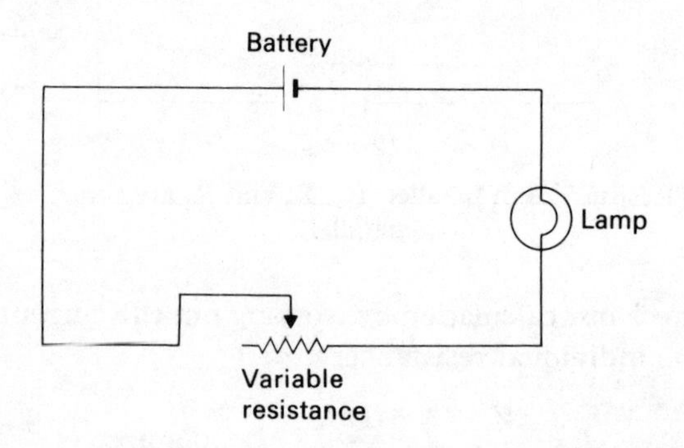

Fig. 7/11. The use of a variable resistance as a light dimmer.

INTERNAL RESISTANCE

If we measure the voltage of a battery when it is not connected in a circuit, and then again when it is supplying current, we find in the latter there is a drop in voltage. This is because all batteries possess *internal resistance*. But, why is this only noticeable when the battery is in a circuit? This is because there will only be a voltage drop when *current is flowing* (through the internal resistance). This voltage drop is proportional to the current flowing and is therefore greater and more noticeable with larger currents. For example, if you try to start a car when the headlights are on, you will notice the lights dim. This is because the starter motor draws a very large current—about 200 amps—and this will produce a large voltage drop across the battery (about 2 volts) which produces the dimming of the lights.

An experiment to show internal resistance is shown in Fig. 7/12. (In this circuit we have represented the internal resistance by an imaginary resistance in between the plates of the battery.)

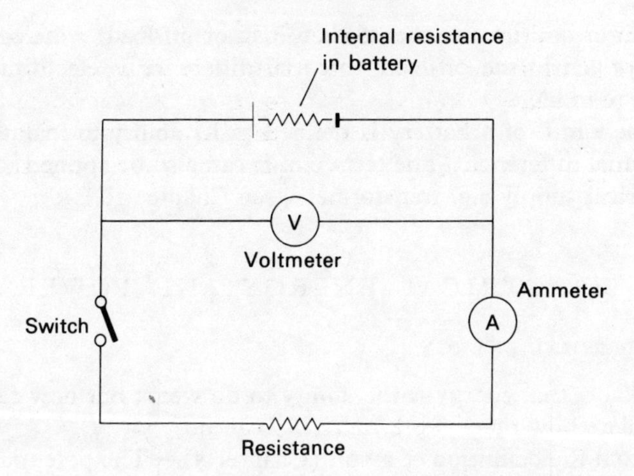

Fig. 7/12. Experiment to illustrate the internal resistance of a battery.

Readings taken:

(a) Voltage—open circuit (switch off) : 6 volts
(b) Voltage—complete circuit (switch on) : $5\frac{1}{2}$ volts
(c) Current—5 amps.

From Ohm's Law:

$$R = \frac{V}{I}$$

$$\text{so} \quad R_{\text{int}} = \frac{V_{\text{drop}}}{I}$$

$$R_{\text{int}} = \frac{\frac{1}{2}}{5}$$

$$R_{\text{int}} = 0.1\ \Omega.$$

Since the internal resistance of most batteries is very small, we may ignore it for most calculations.

ELECTROMOTIVE FORCE

As you see from the above experiment, the potential difference that a battery supplies depends on whether it is part of a complete circuit or not (i.e. whether it is on-load or off-load). The correct term given to the off-load potential difference is electromotive force (e.m.f.).

The e.m.f. of a battery is defined as its ability to maintain a potential difference. (The term e.m.f. can also be applied to any electrical supply e.g. transformers; see Chapter 10.)

ELECTRICAL ENERGY AND POWER

ELECTRICAL ENERGY

We know that energy is the ability to do work; but how can we calculate how much work *electricity* can do?

From the definition of a volt (Chapter 5)—'The potential at a point in volts is equal to the work done in joules in bringing 1 coulomb of positive charge from infinity to that point'—we could say that 'one joule of work is done when 1 coulomb moves through a potential difference of 1 volt', *or* remembering 1 amp for 1 second = 1 coulomb.

One joule of work is done if a PD of 1 volt drives a current of 1 amp for 1 second, for example:

(a) 1 coulomb moving through 10 volts = 10 joules of work;
(b) 3 coulombs moving through 10 volts = 30 joules of work.

Therefore: Work done (in joules) = PD × charge

$$\text{or} \quad \text{Work done} = VQ \text{ joules;}$$

and since energy is the ability to do work,

$$\text{Energy} = VQ \text{ joules.}$$

Remembering that in Chapter 7 we said $Q = IT$ (i.e. charge = current × time), we can now say:

$$\text{Energy} = VIT \text{ joules.}$$

So electrical energy can be calculated by multiplying, i.e. PD (volts) × current (amps) × time (seconds).

ELECTRICAL POWER

We know from Chapter 1 that 'power' is 'rate of doing work' or 'rate of using energy'; in other words 'energy per unit time'. So *power* must be *energy* divided by *time*, i.e.

$$\text{Power} = \frac{\text{energy}}{\text{time}}$$

$$= \frac{VIT}{T}$$

$$\text{Power} = VI.$$

Units of power. Since power is work done per unit time, it follows that the unit of power is 'joules per second'. Another term for a 'joule per second' is a 'watt', i.e.

$$\text{Power} = VI \text{ watts.}$$

Since a watt is quite small, we sometimes use kilowatts, where:

$$1 \text{ kilowatt} = 10^3 \text{ watts.}$$

Units of energy. We said earlier that energy = VIT joules, so a joule is a unit of energy.

But since

$$V \times I = \text{watts},$$

therefore: $\text{Energy (in joules)} = \text{watts} \times \text{seconds}$ (or wattseconds).

When talking about electrical energy we usually use wattseconds instead of joules.

Since a wattsecond is quite small, we sometimes use kilowattseconds or kilowatthours, where:

$$1 \text{ kilowattsecond} = 10^3 \text{ wattseconds;}$$

$$1 \text{ kilowatthour} = 3\,600\,000 \text{ wattseconds.}$$

The kilowatthour is the unit used by the electricity board to calculate electricity bills (1 kilowatthour costs about 3p).

Examples of a kilowatthour are:

(a) 3-kW immersion heater for 20 minutes; (b) 1-kW electric fire for 1 hour; (c) 100-watt light bulb for 10 hours.

SUMMARY

Power is the 'rate of doing work' (watts, or joules per second).
Energy is 'work done' (joules or wattseconds).
Power = current × voltage.
Energy = power × time.

WORKED EXAMPLE

A hair dryer has a resistance of 125 ohms and is operated at a potential difference of 250 volts. Calculate: (a) the current flowing through it; (b) the charge flowing through it in 20 seconds; (c) the power consumption; and (d) the energy it consumes in 2 minutes.

(a) From Ohm's Law: $I = \frac{V}{R}$

$$I = \frac{250}{125}$$

$$I = 2 \text{ amps.}$$

(b) Charge = current × time

$$Q = 2 \times 20$$

$$= 40 \text{ coulombs.}$$

(c) Power = current × voltage

$$= 2 \times 250$$

$$= 500 \text{ watts.}$$

(d) Energy = power × time

$$= 500 \times 2 \times 60 \qquad \text{(time in seconds)}$$

$$= 60\,000 \text{ joules.}$$

HEATING EFFECT OF AN ELECTRIC CURRENT

When a current passes through a conductor energy is used up. This energy appears as *heat* (and sometimes light). The amount of energy can be found by using the equation $E = VIT$ joules.

By using Ohm's Law we can modify this equation;

$$\text{since} \quad V = IR$$
$$\text{then} \quad E = IR \times IT$$
$$\text{or} \quad E = I^2RT \text{ joules}$$

This equation is most important because it shows us that heat is proportional to the square of the current, i.e. $H \propto I^2$. In other words, if the current is doubled there will be four times as much heat produced.

We can use these equations to calculate how much energy an x-ray tube consumes, during an exposure. Let us take a typical radiographic exposure, e.g. 100 kV, 250 mA, 0.2 seconds.

$$\text{Energy} = VIT \text{ joules}$$
$$= 100\,000 \times \frac{250}{1000} \times \frac{2}{10}$$
$$= 5000 \text{ joules.}$$

Since 99% of the energy consumed in an x-ray tube appears as heat, it is not surprising that the x-ray tube becomes very hot in use, because during the exposure it is producing heat at the same rate as 25 electric fires.

A very useful application of the heating effect of an electric current is found in a *fuse*. This is a piece of wire whose melting point is accurately known. If a fuse is incorporated into a circuit, and the current in that circuit reaches a dangerous level, the fuse will melt, thus breaking the circuit and preventing the risk of fire or damage to other components in the circuit.

A FURTHER LOOK AT CAPACITANCE

CHARGING CAPACITORS

In Chapter 6 we talked about and drew graphs of the potential on a capacitor against time. Now let us look at how the value of current flowing into the capacitor (charging current) varies with time in the following circuit (Fig. 7/13).

When the circuit is completed the current starts at a maximum and falls to zero. It falls rapidly at first, and then more slowly,

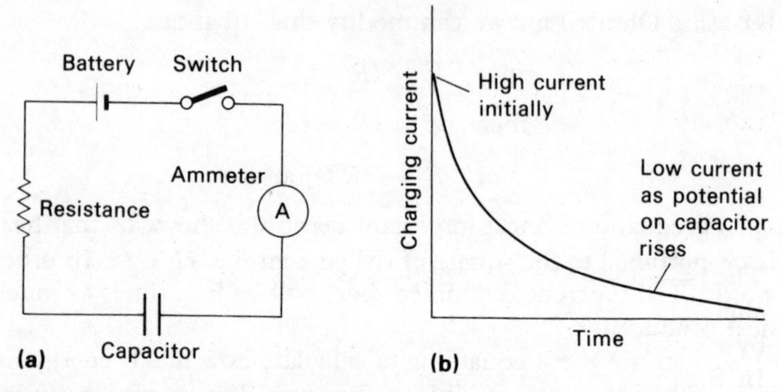

Fig. 7/13. (a) Capacitor charging circuit; (b) graph of charging current against time.

as the potential on the capacitor rises (and therefore tends to oppose the flow of current) until a point is reached where the potential on the capacitor equals that of the battery and there is no further flow of current.

If the capacitor is now removed from the circuit it will retain its charge.

DISCHARGING CAPACITORS

If the capacitor is now connected as shown in Fig. 7/14, and the switch closed the charge it retained will flow around the circuit.

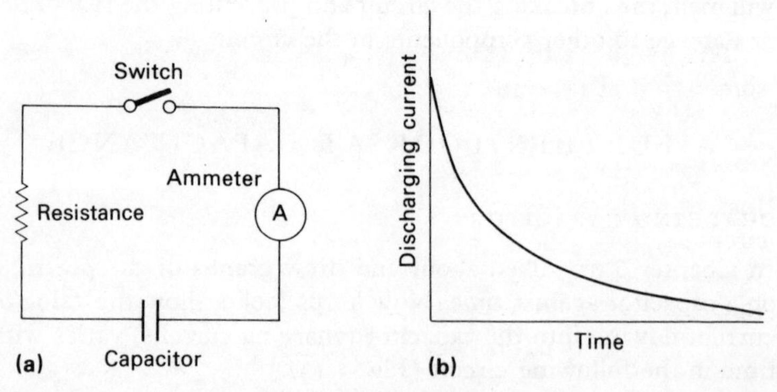

Fig. 7/14. (a) Capacitor discharging circuit; (b) graph of discharging current against time.

If we look at a graph of the discharging current against time, we see that the current starts at a maximum and falls to zero, rapidly at first and then more slowly as the potential on the capacitor decreases.

EFFECT OF RESISTANCE ON CHARGING AND DISCHARGING TIME

The time taken for the capacitor to charge (or discharge) can be altered by varying the value of the charging current, for example by varying the value of the resistance in the circuit.
In Fig. 7/15 we can compare the charging time of a capacitor with (a) a low value of circuit resistance; and (b) a high value of circuit resistance.

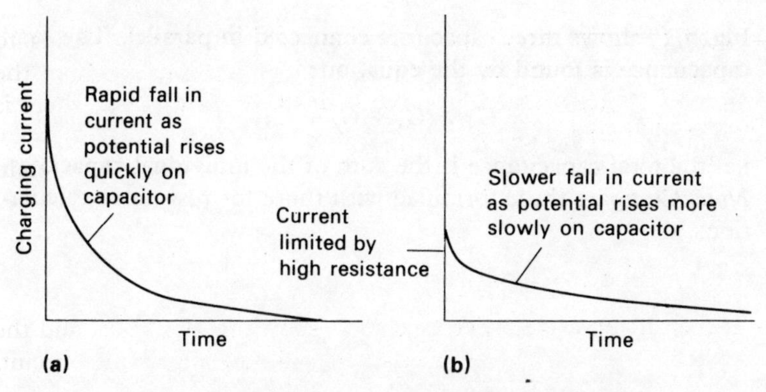

Fig. 7/15. Effect on charging current of low and high resistance. (a) Low resistance; (b) high resistance.

This ability to alter the charging time of a capacitor is used in some types of exposure timers.

COMBINATIONS OF CAPACITORS IN CIRCUITS

Just as with resistances, capacitors can be joined together in circuits.
When capacitors are joined together in series the following equation is used to calculate the total capacitance:

$$\frac{1}{C_T} = \frac{1}{C_1} + \frac{1}{C_2} + \frac{1}{C_3},$$

i.e. the total capacitance is less than any individual capacitor.

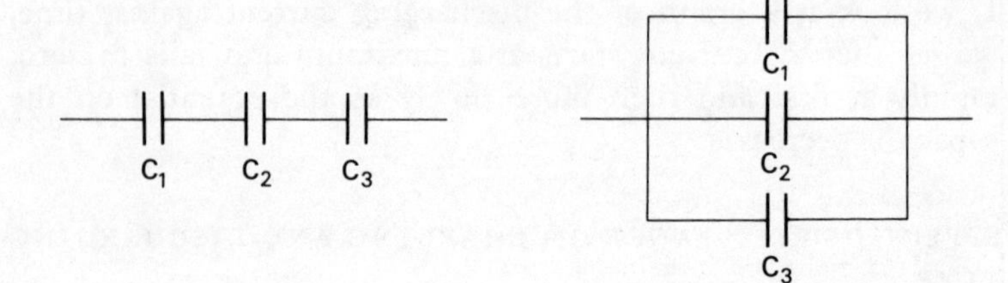

Fig. 7/16. Capacitors in series. Total capacitance = C_T.
$\frac{1}{C_T} = \frac{1}{C_1} + \frac{1}{C_2} + \frac{1}{C_3}$.

Fig. 7/17. Capacitors in parallel. Total capacitance = C_T; $C_T = C_1 + C_2 + C_3$.

Fig. 7/17 shows three capacitors connected in parallel. The total capacitance is found by the equation:

$$C_T = C_1 + C_2 + C_3,$$

i.e. the total capacitance is the sum of the individual capacitors. *Note.* Compare these formulae with those for resistance calculations.

CHAPTER SUMMARY

1. Electric current is the flow of electrons from negative to positive. It is measured in amps (p. 51).
2. Potential difference is a measure of electrical 'pressure'. It is measured in volts (p. 54).
3. Resistance is the opposition to the flow of electric current. It is measured in ohms (p. 54).
4. Ammeters measure current, and they are placed in series with the circuit (p. 54).
5. Voltmeters measure potential difference (p. 54).
6. Ohm's Law states that the current flowing through a conductor is directly proportional to the potential difference between its ends (p. 55).
7. The resistance of a conductor increases with its temperature (p. 58).
8. Internal resistance causes a drop in voltage when current is flowing (p. 64).
9. Power is the rate of doing work (p. 66).
10. Energy is the ability of an object to do work (p. 66).

KEY EQUATIONS

$Q = It$ (charge) (p. 52).

$V = IR$ (voltage) (p. 55).

$R = \frac{sl}{A}$ (resistance) (p. 57).

$R_T = R_1 + R_2 + R_3$ (series resistances; p. 59).

$\frac{1}{R_T} = \frac{1}{R_1} + \frac{1}{R_2} + \frac{1}{R_3}$ (parallel resistances; p. 62).

Power $= VI$ (p. 67).

Energy $= VIt$ (p. 66).

$\frac{1}{C_T} = \frac{1}{C_1} + \frac{1}{C_2} + \frac{1}{C_3}$ (series capacitors, p. 71).

$C_T = C_1 + C_2 + C_3$ (parallel capacitors; p. 72).

Chapter 8
Magnetism

Perhaps the most common example of a magnet is a compass needle. When pivoted the needle always points in a north-south direction. This property of a material is known as magnetism. The end of the compass needle (magnet) which points to the north pole of the earth is known as a *north (seeking) pole* and the end which points to the south pole of the earth a *south pole.*

If we take two magnets and suspend them on threads so that they are free to rotate and bring them towards each other, we find:

1. The north pole of one will *repel* the north pole of the other.
2. The south pole of one will *repel* the south pole of the other.
3. The north pole of one will *attract* the south pole of the other.

This leads us to the **Law of Magnetic Force**, which states that '*like poles repel, unlike poles attract*'. (Notice the similarity to electric charges, the only important difference is that a *magnetic* pole cannot exist on its own.)

Looking back to the first example we used, that of the earth, we realise that the geographic north pole must in fact be a magnetic *south* pole—in order to attract the *north* pole of a compass needle.

WHY ARE SOME MATERIALS MAGNETS, AND OTHERS NOT?

Inside all magnetic materials are domains. We can imagine these as being like small magnets each with a north and a south pole. In non-magnetised materials, all these domains are arranged randomly, thus the total magnetic effect is zero (Fig. 8/1).
In a magnetised material, however, all the domains are lined up, nose-to-tail, i.e. with all their north poles pointing in the same direction, thus reinforcing the magnetic effect (Fig. 8/2).

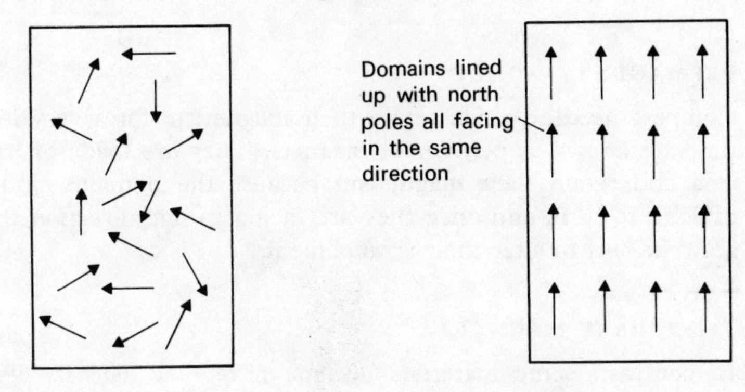

Fig. 8/1. Random arrangement of domains in a non-magnetised material. ↑ indicates north pole.

Fig. 8/2. Arrangement of domains in a magnetised material.

WHAT HAPPENS IF WE TRY TO ISOLATE ONE MAGNETIC POLE?

If we cut a small piece off the end of a magnet to try to isolate the north pole, we find that there is always a south pole associated with it. Thus it is impossible for a magnetic pole to exist on its own.

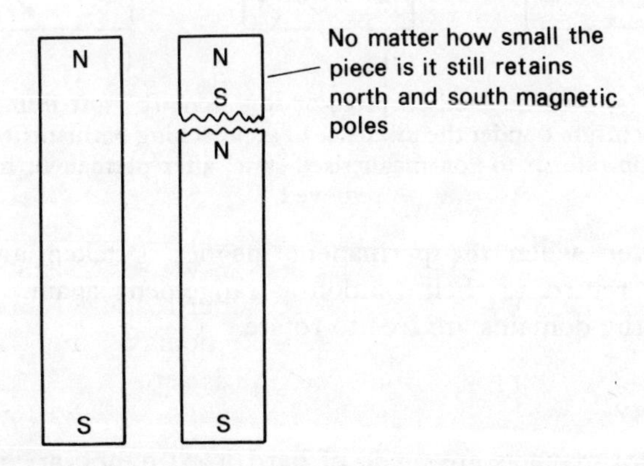

Fig. 8/3. An attempt to isolate the north pole of a magnet.

PERMANENT MAGNETS

Compass needles which retain their magnetism for a very long time are known as permanent magnets. They are made of hard iron and retain their magnetism because the domains find it difficult to turn, and once they are facing in one direction they are reluctant to alter their arrangement.

TEMPORARY MAGNETS

In contrast, some materials become *temporary magnets* when influenced by a permanent magnet. For example if we bring a permanent magnet towards a piece of soft iron, the domains in the iron turn to face the permanent magnet.

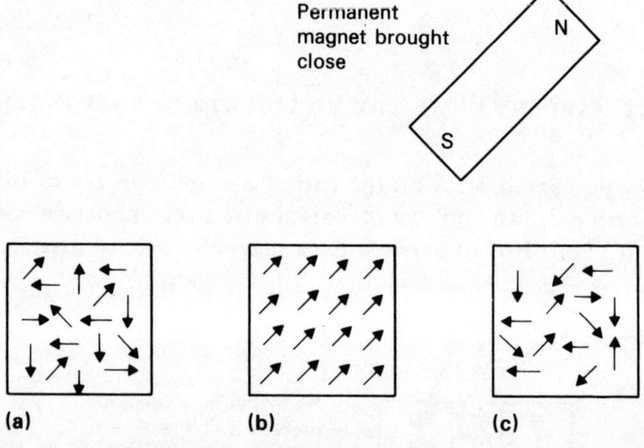

Fig. 8/4. (a) Non-magnetised piece of soft iron; (b) soft iron becomes temporary magnet under the influence of approaching permanent magnet; (c) soft iron returns to non-magnetised state, after permanent magnet is removed.

However, when the permanent magnet is taken away the domains return to their random arrangement again. This is because the domains are free to rotate.

SUMMARY

Permanent magnets are made of hard iron; temporary magnets are made of soft iron.

MAGNETIC FIELDS

Just as electric charges are surrounded by an electric field, so magnets are surrounded by a magnetic field. A magnetic field is an area around a magnet in which a magnetic effect is noticed. Magnetic fields are made up of lines of force. A line of force can be defined as the path taken by an independent north pole moving from the north pole of a magnet to the south.

Lines of force can be observed in two ways:

1. A bar magnet is placed under a sheet of paper and lots of small particles of soft iron (iron filings) are sprinkled on top. When the paper is tapped, the iron filings line up in the paths of the lines of force.

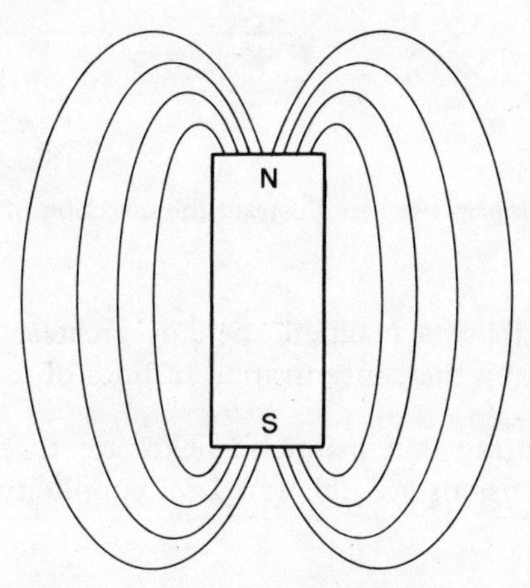

Fig. 8/5. Magnetic field around a bar magnet.

2. The lines of force can be plotted, as above, but by using small compass needles. This method also illustrates their direction (Fig. 8/6).

OBSERVATIONS

1. A line of force originates at the north pole and ends at the south pole.

2. No two lines of force cross each other.

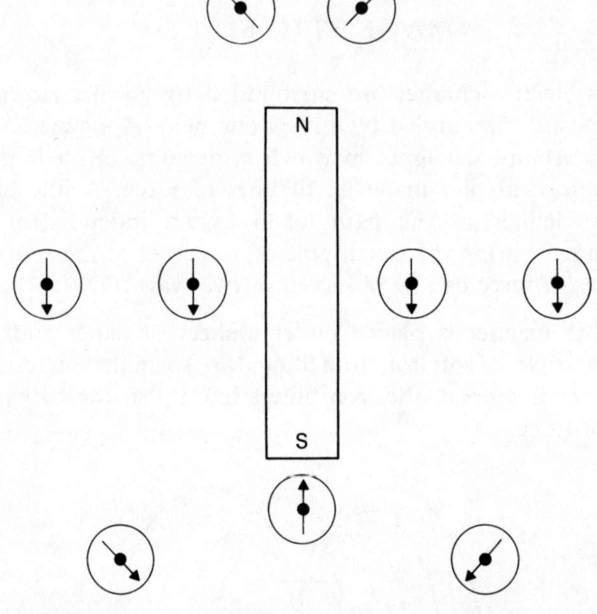

Fig. 8/6. Compass used to illustrate the direction of lines of force.

3. The strength of a magnetic field is greatest near its poles. This is shown by the concentration of lines of force.

It is worth noting that magnetic fields are three-dimensional; only two dimensions are illustrated for simplicity.

OTHER SHAPES OF MAGNET

A horse-shoe magnet is used in the construction of x-ray equipment.
This is really just a bar magnet that has been bent into a horse-shoe shape. It does however offer two important advantages over a conventional bar magnet:

1. The strength of the magnetic field is greater because the poles are closer together.
2. By adding soft iron pole pieces we can produce a magnetic field in which most of the lines of force are parallel to each other.

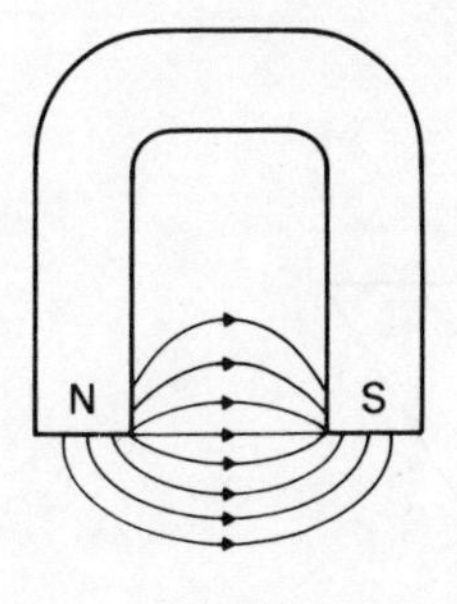

Fig. 8/7. Horse-shoe magnet.

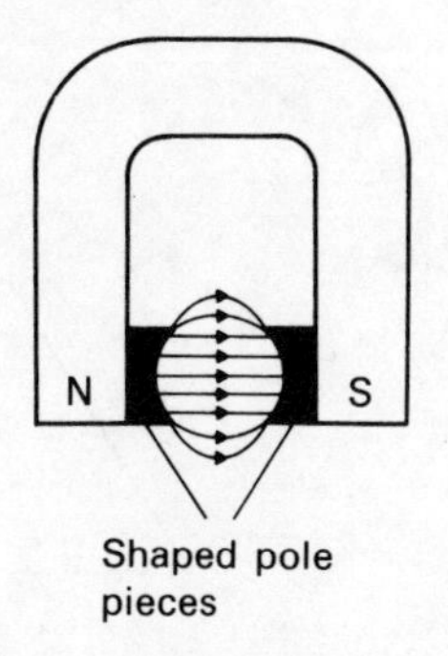

Fig. 8/8. Magnetic field and horse-shoe magnets.

This effect will be used later when we look at the construction of meters (Chapter 8).

MAGNETIC POLE STRENGTH

It is sometimes useful to know how the force of repulsion (or attraction) varies with distance. The equation is similar to that for electric charges;

$$F \propto \frac{m_1 m_2}{d^2}$$

where m_1 and m_2 are the pole strengths.
It is important to notice from the equation that the force is inversely proportional to the *square* of the distance. In other words if we double the distance between the poles, the force is one-quarter of the original.

The material between the poles also affects the force between them. This is known as *permeability*.

MAGNETIC EFFECT OF AN ELECTRIC CURRENT

In 1820 Oersted discovered that there was a magnetic effect associated with the passage of an electric current. When a compass needle is placed near a current-carrying conductor a deflection is noticed when the current is switched on.

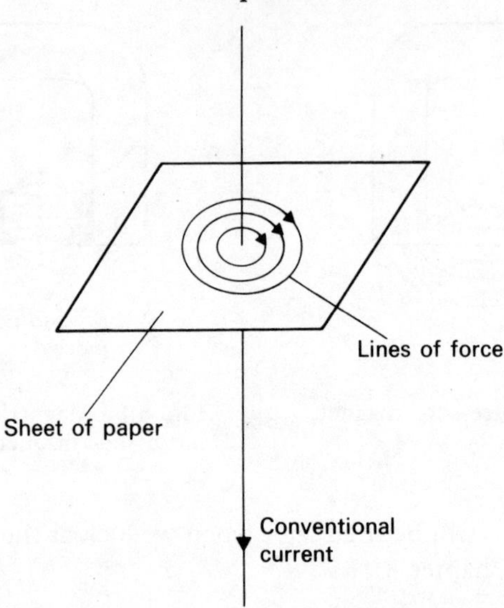

Fig. 8/9. Magnetic field around a current-carrying conductor.

OBSERVATIONS

1. The magnetic field is only present when the current is switched on.
2. The lines of force around the conductor are circular.
3. Their direction (clockwise or anticlockwise) depends on the direction of the current.

The direction of the lines of force can be predicted by using Maxwell's Corkscrew Rule (which applies to conventional current). If the current is *in* to the paper, the magnetic field is *clockwise* and if the current comes *out* of the paper, the magnetic field is *anticlockwise*. Compare this with screwing a corkscrew into a cork; when the corkscrew is going into the cork, the direction of rotation is clockwise (Fig. 8/10).
The dot (·) in the centre indicates that the current is coming out towards us, and the cross (+) indicates that the current is going away from us. The magnetic fields produced by this method are, however, too weak to be of any practical use.

If the conductor is wound into the shape of a coil (solenoid) we find that the magnetic effect is much stronger. This is because each of the turns produces a magnetic field which reinforces the next (Figs. 8/11 & 8/12).
(Note the similarity to a bar magnet.)

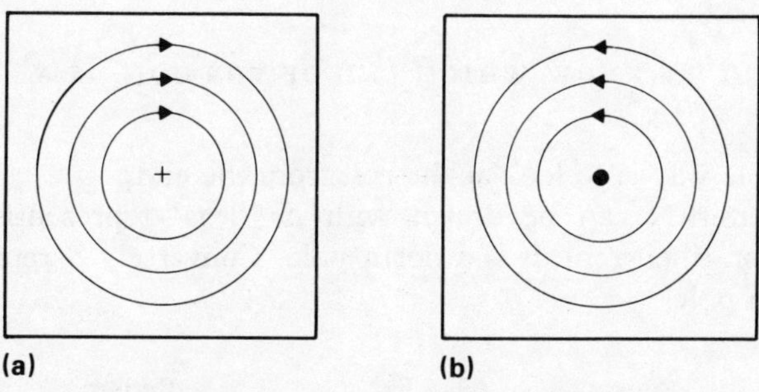

Fig. 8/10. Direction of lines of force around a straight current-carrying conductor. (a) Current *in* to page (away from us); (b) current *out* of page (towards us).

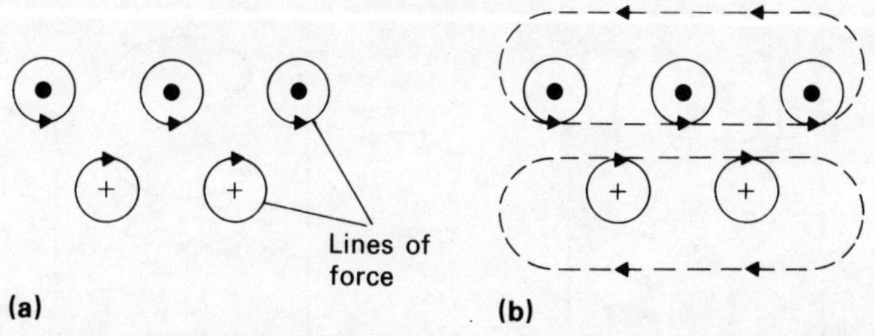

Fig. 8/11. Lines of force around the turns of a current-carrying coil. (a) Cross-section through coil; (b) the combined effect of the lines of force.

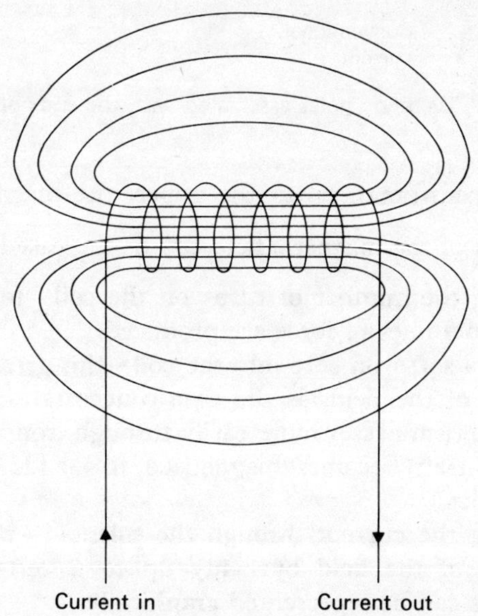

Fig. 8/12. The resultant magnetic field around the coil.

HOW DO WE KNOW WHICH END OF THE COIL IS A NORTH POLE?

A simple way is to look at the coil from the end;
If a letter N can be drawn with its 'legs' representing the direction of current, it is a north pole. Conversely S represents a south pole.

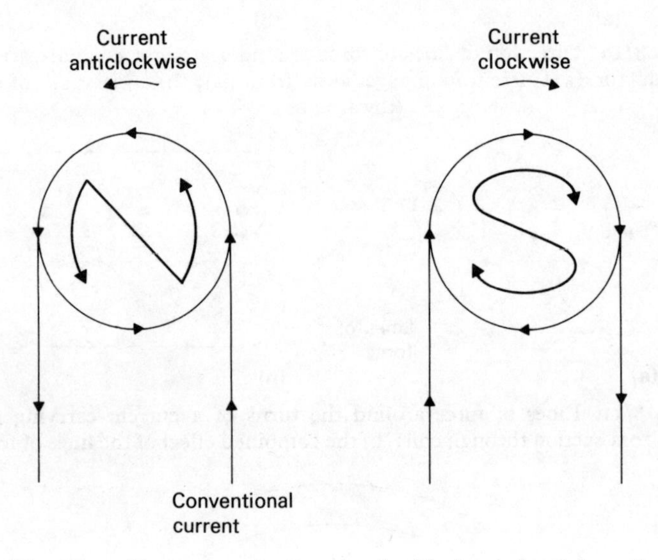

Fig. 8/13. Magnetic poles associated with the ends of the coil.

STRENGTH OF FIELD

The strength of the magnetic field can be increased by:

1. Increasing the number of turns on the coil—for each extra turn, additional lines of force are produced.
2. Inserting a soft iron core into the coil—this greatly increases the strength of the field, as the core concentrates the lines of force. (Magnetism passes more easily through iron than air, and as the iron itself becomes magnetised this adds to the total magnetic effect.)
3. Increasing the current through the solenoid—this increases the strength of the field but only up to a certain value of current. This can be represented graphically.

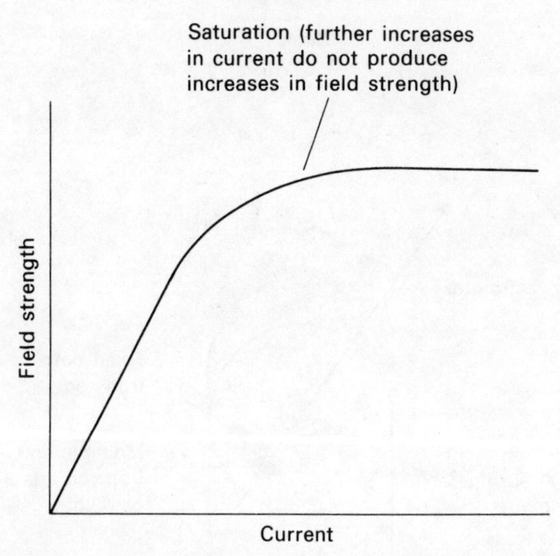

Fig. 8/14. Effect on field strength when the volume of current is increased.

When saturation is reached, further increases in current do *not* produce any further increases in field strength. At this point all the domains are lined up in the same direction.

A magnet of the type we have just talked about (solenoid with soft iron core) is known as an electromagnet.

The soft iron core acts as a temporary magnet, and therefore does not retain much of its magnetism after the current has been switched off. Thus we have a magnet which can be switched on and off as required.

PRACTICAL APPLICATIONS

1. Equipment for removing metallic foreign bodies from the eye.
2. Brakes on x-ray equipment e.g. to hold x-ray tubes in position.
3. Remotely operated electric switches, known as relays.

MOVING IRON METERS

We have seen that the passage of an electric current produces a magnetic field, and that the strength of this field is proportional to the current which produces it. If therefore we could measure

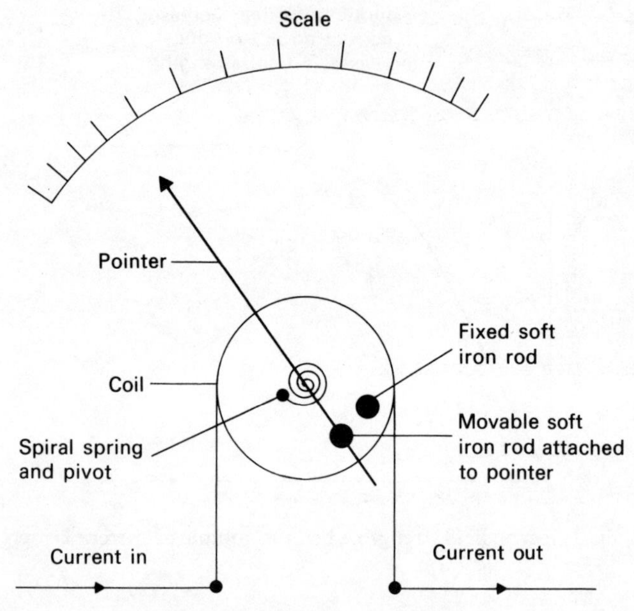

Fig. 8/15. Moving iron meter.

the strength of the magnetic effect, this would be proportional to the current flowing, i.e. we have a means of measuring electric current.

PARTS

The meter consists of a coil of wire joined to two terminals. Inside the coil are two soft iron rods, one of which is fixed, the other free to move. The moveable rod is attached through a pivot, to a pointer which moves across a scale.

OPERATION

When current passes through the solenoid, both soft iron rods become magnetised, with the same polarity. Therefore they repel each other (like poles repel) and the one which is pivoted moves away from the fixed rod.

The force which moves them is proportional to the current flowing. The turning movement is opposed by the spiral spring, which brings the pointer back to zero when the current is switched off.

The scale on the meter is said to be non-linear, i.e. the graduations are not equally spaced. This is because the force between two magnets is inversely proportional to the square of the distance separating them (see p. 79).

ADVANTAGES

One of the big advantages of this instrument is that it will work equally well if the current flows in either direction, since both soft iron rods will receive the same polarity and will therefore always repel each other. These meters are relatively cheap to produce, and are robust.

DISADVANTAGES

The main disadvantage is that the non-linear scale can produce inaccurate readings, particularly at the upper and lower ends of the scale.

FORCE ON A CURRENT-CARRYING CONDUCTOR IN A MAGNETIC FIELD

Since we know that there are always forces of attraction or repulsion between magnets (magnetic fields), it would be reasonable to assume that such a force would exist between a permanent magnet and a current-carrying conductor.

This in fact is the case and it can be demonstrated as follows (Fig. 8/16).

When a current flows through the conductor it moves at right angles to the magnetic field, in this example downwards. (There is *no* movement towards either of the poles.)

If the direction of current is reversed, we find that the movement of the conductor is reversed, i.e. upwards. Similarly, if the direction of the magnetic field is reversed, the movement is again reversed.

The rule which determines the direction of motion of the conductor is *Fleming's Left Hand Rule* (Fig. 8/18). This states that 'if the first three fingers of the left hand are extended at right angles to each other, so that the first finger indicates the direction of the magnetic field, the second finger indicates the direction of

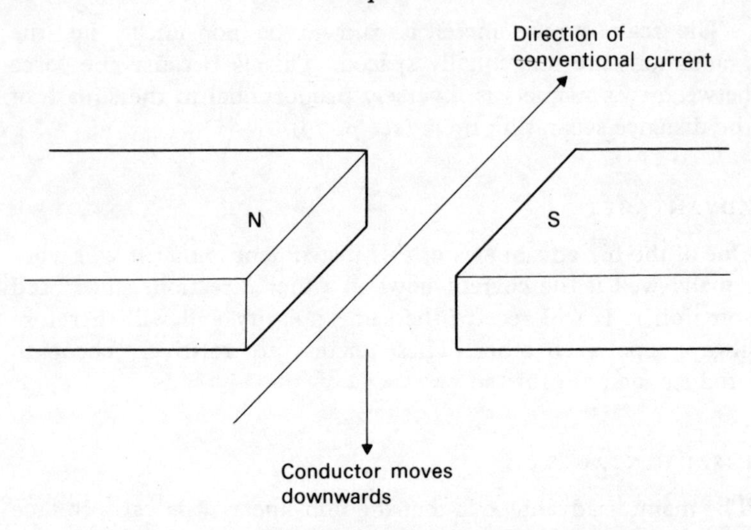

Fig. 8/16. Force on a current-carrying conductor in a magnetic field.

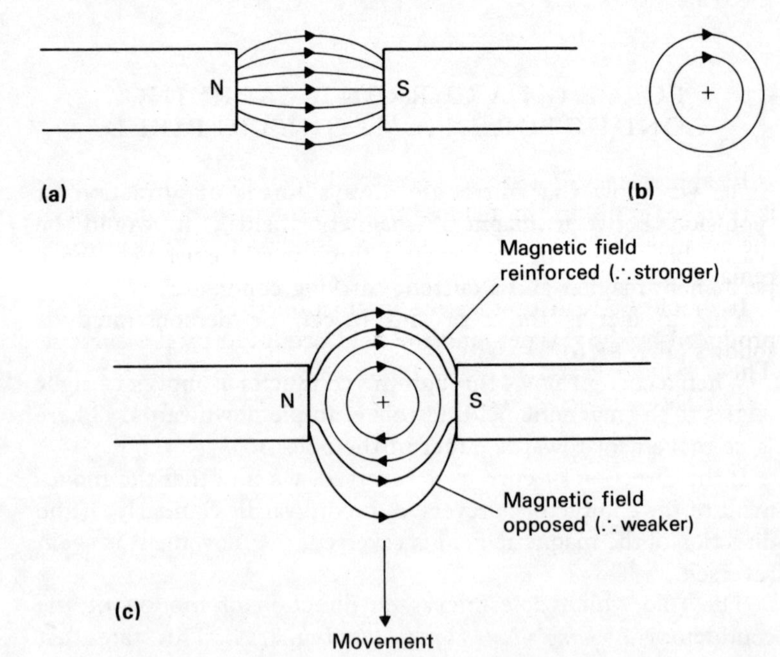

Fig. 8/17. Reason for this movement. (a) Magnetic field produced by permanent magnets; (b) magnetic field produced by current-carrying conductor; (c) result of interaction of these two magnetic fields.

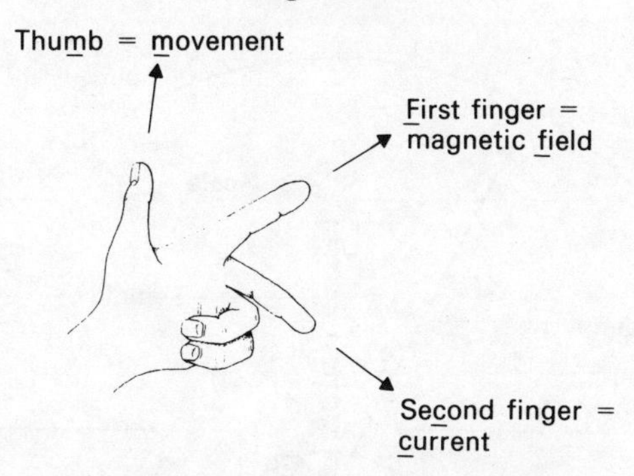

Fig. 8/18. Fleming's Left Hand Rule.

the current, then the thumb will indicate the direction of movement of the conductor'. An easy way to remember this is:

First finger = Field
Second finger = Current
Thumb = Movement.

By applying this rule we can see why the direction of movement is reversed if either the field or the current is reversed. If both the field and the current are reversed, the direction of movement remains the same.

It should be realised that the interaction is between the field produced by the magnet, and the field produced by the current. The conductor plays no part except as a carrier of the current.

The above effect is known as the *motor* effect.

APPLICATION OF THE MOTOR EFFECT

A most widely used instrument in x-ray equipment is the moving coil meter.

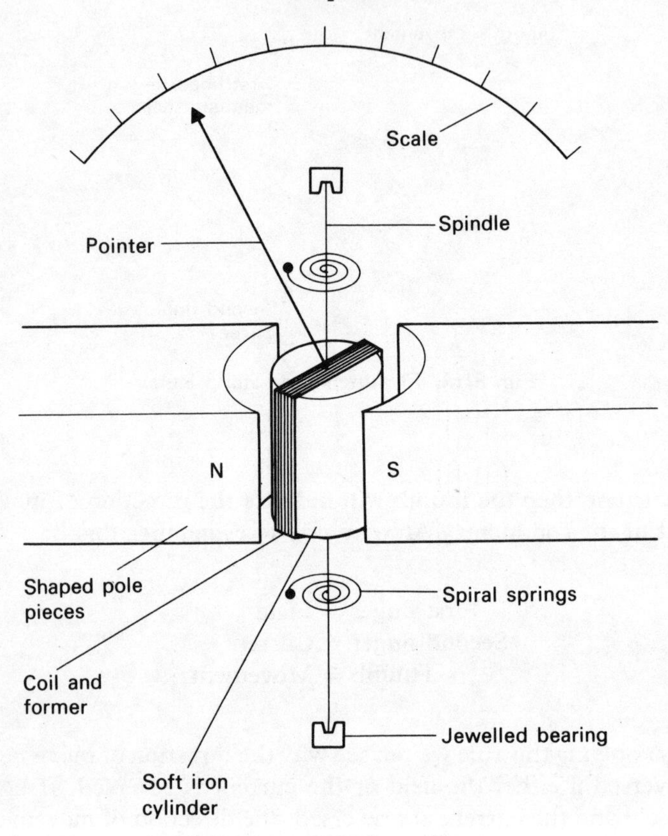

Fig. 8/19. (a) Moving coil meter.

PARTS

1. Horse-shoe shaped magnet with curved pole pieces, with a cylindrical soft iron core.
2. Aluminium former (frame) with a coil of thin (insulated) wire wound on it.
3. A spindle is attached to the former, and held in place with spiral springs and jewelled bearings.
4. A pointer is attached to the former, and moves over a linear scale.

PRINCIPLE OF OPERATION

The principle depends on the motor effect.

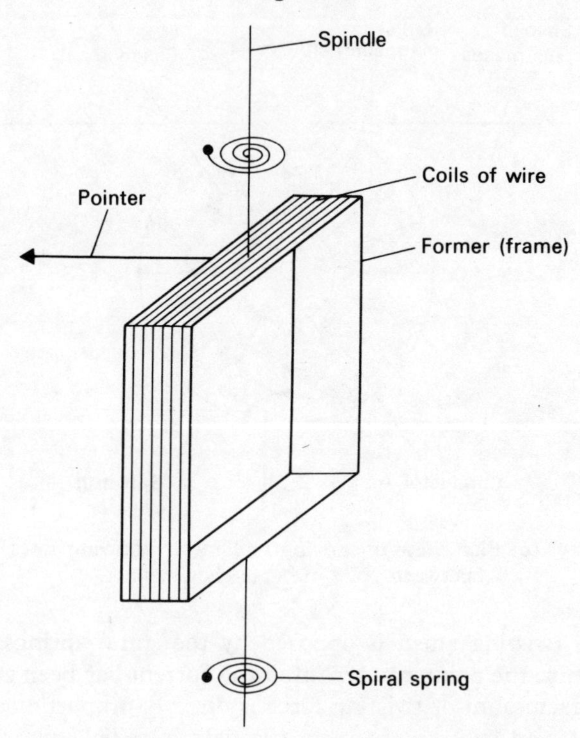

Fig. 8/19. (b) Moving parts of moving coil meter.

It is important to remember that if the conductor is parallel to the magnetic field, it will experience no force.

Looking at Fig. 8/19c, we can assume that the coil is made up of a single turn of wire (in practice this is not normally the case).

1. Looking at wire A. This is carrying (conventional) current up, towards us, and the magnetic field is left to right. Applying Fleming's Left Hand Rule, we find the resulting movement is upwards.

2. Looking at wire B. This is carrying current down, away from us, and the magnetic field is from left to right.

Applying Fleming's Left Hand Rule, we find the resulting movement is downwards.

The combined effect of these two forces causes the coil and former to twist (clockwise) about the pivot.

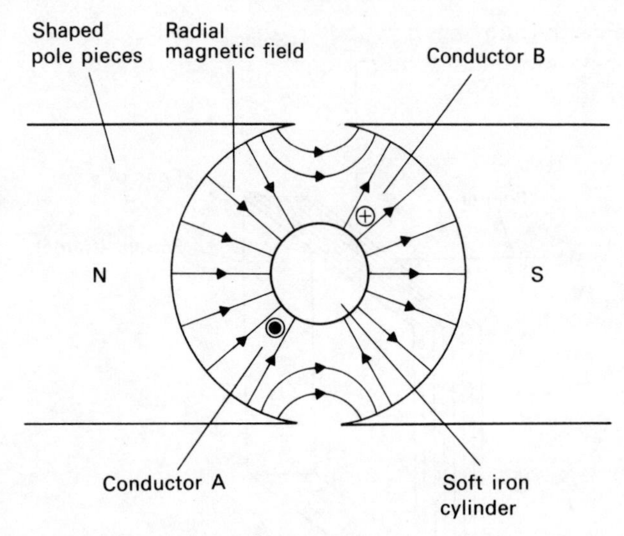

Fig. 8/19. (c) Plan view of moving coil meter showing lines of force between poles of horse-shoe magnet.

This twisting effect is opposed by the spiral springs, which also return the pointer to zero after the current has been switched off. The amount of twisting force is directly proportional to the current, and because the magnetic field is radial, with lines of force equally spaced, equal amounts of current will produce equal increases in the twisting force. We can see from this that the graduations on the scale will be equally spaced, i.e. a linear scale.

FUNCTION OF PARTS

The shaped pole pieces, soft iron cylinder, spiral springs and the former serve certain functions:

1. The curved pole pieces produce lines of force which are parallel to each other and equally spaced. The soft iron cylinder changes the parallel magnetic field into a radial field, i.e. one in which the lines of force are always at right angles to the movement of the coil and former.

2. The spiral springs, as well as returning the pointer to zero after the current is switched off, also lead the current in and out of the coil.

3. The former helps the pointer settle quickly at the correct reading. How it does this will be explained in Chapter 9.

Unlike the moving iron meter, the direction of the current must be known, since reversing the current would reverse the movement of the pointer.

RESISTANCE OF THE METER

As ammeters are connected in series with the circuit, it is important that their resistance is very small, so that they do not alter the current flowing through the circuit.

ADVANTAGES

1. Accurate over the whole scale.
2. Very sensitive instrument (can be made to measure very small currents and is then known as a galvanometer).

DISADVANTAGES

1. Expensive to make.
2. Fragile.
3. Measures current in one direction only.

This instrument is usually made to measure mA, but it can easily be modified to measure larger currents, by connecting low resistances in parallel with it so that a known proportion passes through the meter, the rest is diverted around it.

An extension of the principle of the moving coil meter is found in some electric motors.

CONVERTING AMMETERS TO VOLTMETERS

Moving coil meters can also be modified to measure voltage (potential difference). In this case they are connected in parallel with the part of the circuit across which the voltage is queried. It is important that the voltmeter does not interfere with the current that may be flowing through the circuit at the time. To ensure this, a very high resistance is included in series with the meter so that the current flowing through the meter will be very small, i.e. very little effect on the circuit.

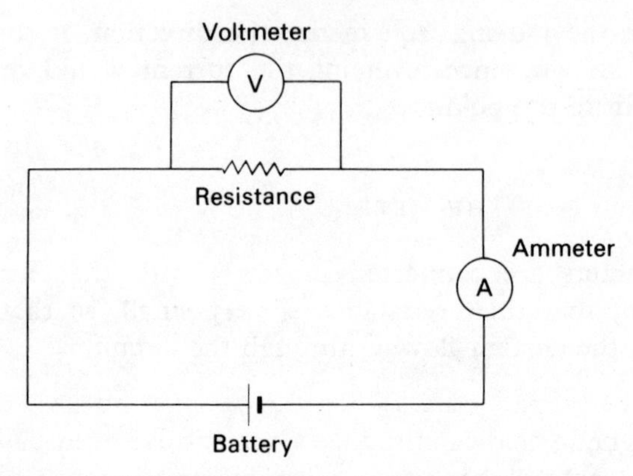

Fig. 8/20. Moving coil meters used as ammeter and voltmeters.

The principle of operation of the moving coil meter used to measure voltage is found in Ohm's Law ($V \propto I$). In other words the meter is made to pass a small current which is proportional to the voltage being measured. The scale of the instrument is calibrated in volts.

CHAPTER SUMMARY

1. Like poles repel, unlike poles attract (p. 74).
2. Permanent magnets are made of hard iron (p. 76).
3. Temporary magnets are made of soft iron (p. 76).
4. A magnetic field is the area around a magnet in which a magnetic effect is noticed. It is made up of lines of force travelling from the north pole to the south (p. 77).
5. There is a magnetic effect associated with the passage of an electric current. Moving iron meters make use of this effect to measure the value of currents (pp. 79–83).
6. A current carrying conductor experiences a force when it is placed in a magnetic field. The direction of the force is found by Fleming's Left Hand Rule. Moving coil meters use this effect to measure the value of electric currents (pp. 85–86).
7. Ammeters can be modified to measure voltage by placing a high resistance in series with the meter (p. 91).

Chapter 9
Electromagnetic induction

The English physicist Faraday was convinced that since a current produced a magnetic field, then the opposite must also be true, i.e. a magnetic field can produce a current. He eventually proved this and showed that when a conductor is moved through a magnetic field, an e.m.f. is induced in the conductor. If the ends of the conductor are joined to form a complete circuit then current will flow.

The experiment to illustrate this is similar to that for the motor effect, i.e. the movement of the conductor must not be parallel to the magnetic field.

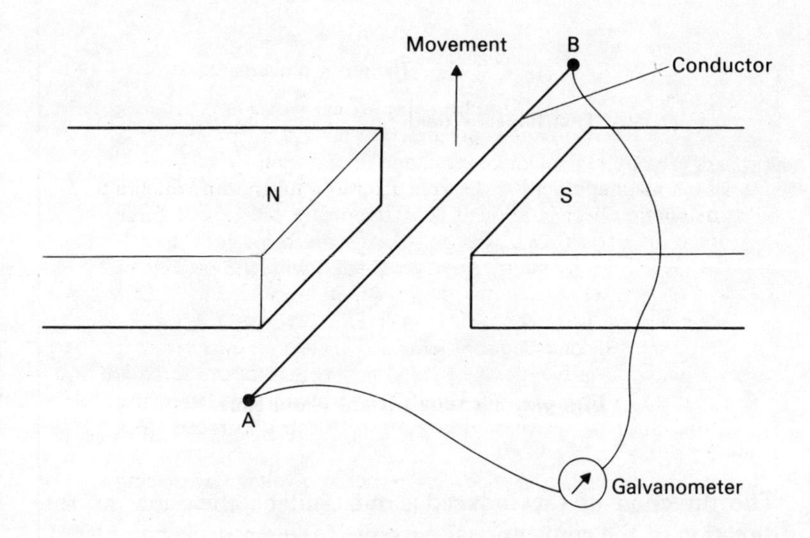

Fig. 9/1. Production of an e.m.f. by movement of a conductor through a magnetic field.

FURTHER FINDINGS

1. If the direction of movement, or field is reversed, the direction of the induced e.m.f., is also reversed.
2. The magnitude of the induced e.m.f. is proportional to the speed of movement and strength of the magnetic field.
3. The effect is the same if the magnet is moved and the conductor remains stationary.
4. No e.m.f. is induced unless there is movement between the conductor and the magnet.

The rule which tells us the direction of the induced e.m.f. is *Fleming's Right Hand Rule* which states that 'if the first three fingers of the right hand are extended at right angles to each other so that the thumb indicates the direction of movement, the first finger indicates the direction of the magnetic field, then the second finger indicates the direction of the induced e.m.f.'.

Thumb = Movement
First finger = Field
Second finger = E.m.f.

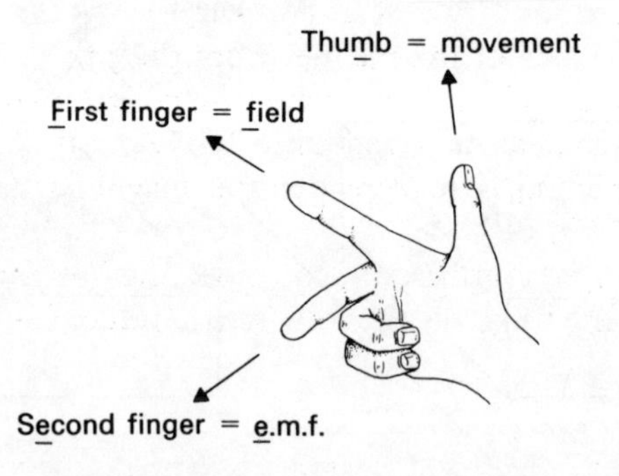

Fig. 9/2. Fleming's Right Hand Rule.

The direction of the induced e.m.f. will be the same as the direction of the conventional current if the circuit is completed. Thus, in Fig. 9/1, if the conductor is moved upwards the current will flow from A to B.

The e.m.f. induced in a straight conductor is too small to be of any practical value, but its value can be increased by winding the wire into a coil. If the coil is connected to a very sensitive milliammeter (galvanometer) and a bar magnet is moved towards or away from one end of the coil a much stronger current is detected.

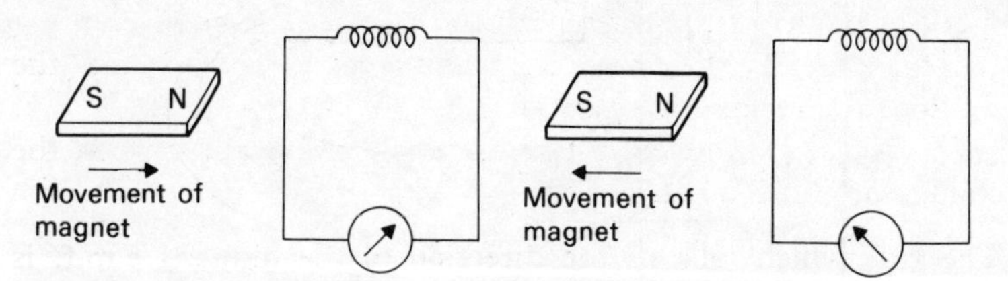

Fig. 9/3. Generation of an emf using permanent magnet and coil.

OBSERVATIONS

1. The *direction* of the induced e.m.f. depends on:
(a) which pole is nearer the coil;
(b) whether the pole is moving towards or away from the coil.

2. The *magnitude* of the induced e.m.f. depends on:
(a) the speed of movement;
(b) the strength of the magnet;
(c) the number of turns on the coil;
(d) the distance of the magnet from the coil.

3. An e.m.f. is only induced when there is relative movement between the magnet and the coil. The easiest way to imagine what is happening is to consider the magnetic field around a bar magnet (Fig. 9/5).
As it is brought towards the coil, the lines of force from the magnet 'cut' the turns on the coil, which induces an e.m.f. across them.

OPPOSITION TO THE FLOW OF CURRENT

Let us refer back to our straight conductor, the current that we are inducing in our conductor must also be producing its own magnetic field! As we already know, this field will produce a force on the conductor.

So can we work out which direction this force will be? Will it oppose the movement we are putting in, or will it add to it?

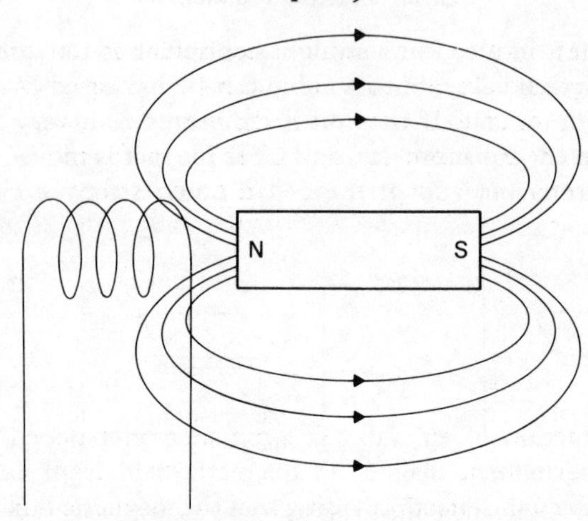

Fig. 9/4. Lines of force around permanent magnet 'cutting' the turns on the coil.

In Fig. 9/1 if we move the conductor upwards, from Fleming's Right Hand Rule we can calculate that the current will flow from A to B. By applying Fleming's Left Hand Rule, to find the effect of this induced current, we find that the direction of force is downwards, i.e. there is a force trying to prevent the movement of the conductor.

All these observations lead us to the *Laws of Electromagnetic Induction.*

Law 1. Whenever the number of lines of force linked with a conductor changes, an e.m.f. is induced in the conductor.
Law 2. The induced e.m.f. is proportional to the rate of cutting of lines of force.
Law 3. Lenz's Law. The induced e.m.f. is in such a direction as to oppose the movement producing it.

Because this induced e.m.f. is in the opposite direction to the e.m.f. producing the movement, it is called 'back e.m.f.'. An important application of Lenz's Law is found in the moving coil meter. We said that the coil was wound on an aluminium former, so that when the coil moves so does the former. Current is therefore induced to flow in the former when the coil moves and from Lenz's Law we know that the direction of this induced current is such as to oppose the motion causing it.

The effect of this is to reduce the oscillations of the pointer enabling quick accurate readings without having to wait a long time for the needle to stop swinging; and since current is only induced in the former when the coil is moving, there is no effect on the pointer once it has stopped at its final reading.

This effect is called *electromagnetic damping or eddy-current damping*.

MUTUAL INDUCTION

The production of an induced e.m.f. does not depend on the use of a permanent magnet. A magnetic field from any source will have the same effect providing that the magnetic flux (number of lines of force) linked with the conductor can be made to vary.

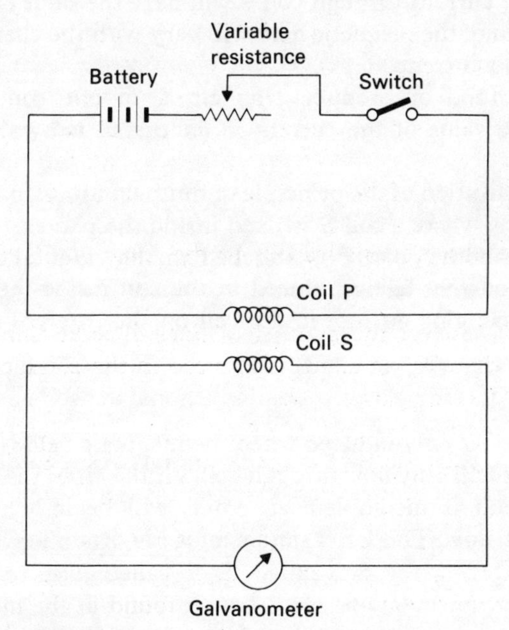

Fig. 9/5. Experiment to demonstrate electromagnetic induction.

In Fig. 9/5 we can see that the permanent magnet used previously has been replaced by a circuit comprising, a battery, variable resistance, switch, and a coil of wire. The rest of the circuit is the same as before, i.e. a coil of wire S connected through a galvanometer.

When the upper part of the circuit is completed by closing the switch, a momentary deflection will be seen on the galvanometer. This is because the current in coil P *together with the magnetic field* is rising from zero and cutting coil S. This changing magnetic field induces an e.m.f. in coil S and current flows through the galvanometer.

When the current in coil P has reached a steady value induction ceases since the magnetic field is now steady (i.e. no change in magnetic flux linkage) and the galvanometer returns to zero.

If the current in coil P is now switched off, the magnetic field will collapse causing another momentary deflection on the galvanometer, but this time the deflection is in the opposite direction (because of the change of direction of the magnetic field).

We have only described switching on and off the current, but varying the current through coil P will have the same effect since the strength of the magnetic field will vary with the change in the value of the current.

This method of producing an e.m.f. in one conductor by varying the value of the current in another is known as *mutual induction.*

One application of the principle of mutual induction is in heart pacemakers, where a coil S is fixed inside the patient (under the skin) and another coil P is attached to the external surface of the chest, current being induced in the coil inside the patient's chest by changing current in the coil on the external surface of the patient.

NOTE

1. An e.m.f. is only induced when the magnetic field is *changing.*
2. Current will only flow through coil S if the circuit is complete. If the circuit is incomplete an e.m.f. will be induced but no current will flow. This latter statement is important in Chapter 10.

SOME CONCLUSIONS

1. The direction of the induced e.m.f. is seen to depend on whether the magnetic field is rising or falling and also on the direction of the current in the primary circuit.
2. The maximum value of the induced e.m.f. will be increased if the two windings are placed closer together. The maximum

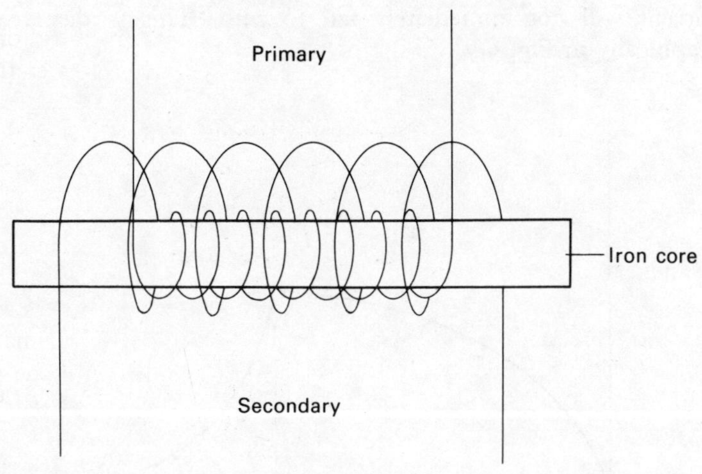

Fig. 9/6. Primary and secondary coils wound on top of each other on an iron core.

possible being if the two windings are wound on top of each other.

3. By inserting a soft iron core into the coils, the value of the induced e.m.f. can be further increased.

4. The maximum value of the induced e.m.f. can also be raised by increasing the number of turns on the secondary winding (the magnetic field will cut more turns of wire).

(That part of the circuit which contains the battery and coil is known as the *primary* circuit and is said to carry the *primary* current. That part of the circuit in which current is induced, i.e. coil S and the galvanometer, is known as the *secondary* circuit.)

SELF-INDUCTION

We have just stated that an e.m.f. will be induced in any conductor which is in a changing magnetic field.

The magnetic field that the primary is producing cuts *its own windings* as well as the secondary. This will also induce an e.m.f. in the primary. This effect is called *self-induction.*

We know from Lenz's Law that this induced e.m.f. will oppose the current producing it. It will therefore tend to oppose the rise in current and conversely try to maintain the current when it is falling, i.e. instead of the current in the primary circuit rising immediately when the switch is closed, it will rise only slowly to its maximum value. Similarly, when the switch is opened the current will not immediately fall to zero. This is displayed graphically in Fig. 9/7.

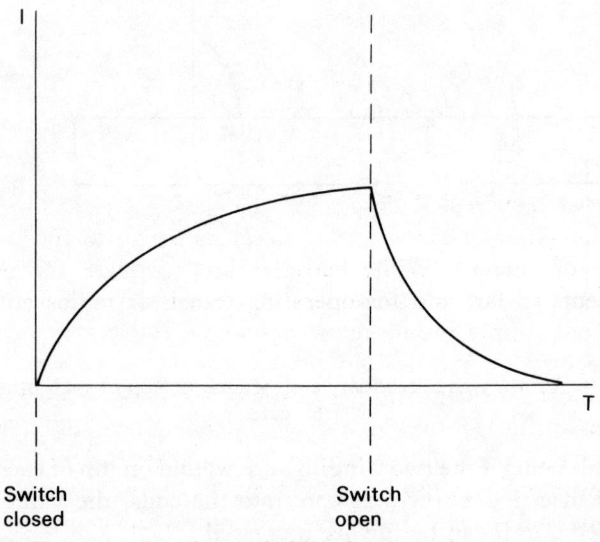

Fig. 9/7. Delay in the rise and fall of value of current caused by back e.m.f.

CHAPTER SUMMARY

1. When there is relative movement between a conductor and a magnetic field an e.m.f. is induced in the conductor (p. 93).
2. The direction of the induced e.m.f. is found by Fleming's Right Hand Rule (p. 94).
3. The direction of the induced e.m.f. is such as to oppose the motion producing it (p. 96).
4. Mutual induction is the effect a coil experiences due to changes in the magnetic field from another coil (p. 97).
5. Self-induction is the effect a coil experiences due to changes in its own magnetic field (p. 99).
6. Mutual and self-induction only occur when there is a change in magnetic flux linkage (p. 98).

Chapter 10
Alternating current

So far we have talked about electric currents which flow in one direction only, i.e. *direct current* (d.c.) and a battery has been the source of current. While batteries are adequate for our experiments so far, and for operating transistor radios etc., they could not supply the needs of the whole country for purposes such as cooking and heating.

So, most of our electricity is produced in electricity-generating stations using the principle of electromagnetic induction. The simplest and most efficient type of generator is called an alternator. This produces a current which changes (alternates) in direction 100 times per second. This type of current is known as *alternating current* (a.c.).

Alternating current is used throughout the world for domestic and industrial uses, offering important advantages over direct current.

ADVANTAGES OF ALTERNATING CURRENT

1. Easier and cheaper to produce than direct current.
2. The value of the voltage of alternating current can be changed easily, using a device known as a transformer. Transformers only work on alternating current.

GENERATION OF ALTERNATING CURRENT

To summarise what we said in Chapter 9: 'when there is relative movement between a conductor and a magnetic field, so that the conductor cuts the lines of force, an e.m.f. will be induced in the conductor'.

This effect was apparent when either (a) the conductor moves and the magnet is stationary; (b) the magnet moves and the conductor is stationary; or (c) the conductor and the magnet are stationary and the strength of the magnetic field varies.

All of these methods are used in different types of electrical equipment but in this unit we will only consider (a) i.e. the conductor moving in the stationary magnetic field.

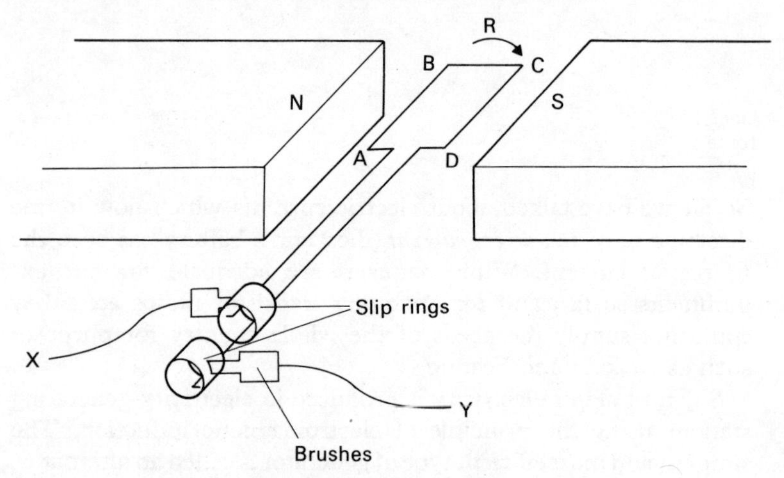

Fig. 10/1. A.C. generator.

Let us consider the following set-up:
N and S are the north and south poles respectively of a horse-shoe magnet; ABCD is a single coil of wire pivoted on a central axis and R is indicating the direction of rotation of the coil. Slip rings and brushes are a means of keeping the connections X and Y in contact with the coil at all times (in other words they prevent the wires getting tangled up as the coil rotates).

Now let us look at what happens when the coil is rotated in the direction R. Applying Fleming's Right Hand Rule:

1. Side AB moves up, therefore current flows from A to B.
2. Side CD moves down, therefore current flows from C to D.

There is no effect on BC since movement is parallel to the magnetic field. Therefore the current flows through the circuit X–A–B–C–D–Y.

Now what happens when the coil has rotated through 180°, i.e. side AB is where side CD was and vice versa:

1. Side CD moves up, therefore the current flows from D to C.
2. Side AB moves down, therefore the current flows from B to A.

Thus the current flows through the circuit Y–D–C–B–A–X.

Looking at the coil from above we can plot a graph of the direction of the current compared with the position of the coil at any moment in time, starting with the coil in the vertical position.

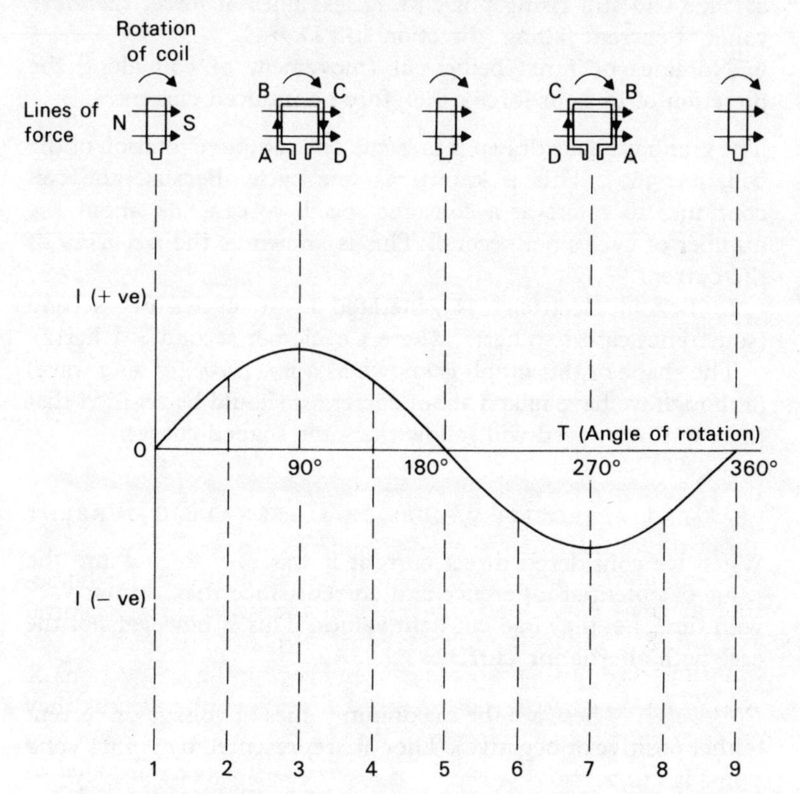

Fig. 10/2. How the current varies with the rotation of the coil.

SUMMARY OF GRAPH (FIG. 10/2)

1. No lines of force being cut (movement of coil along the direction of lines of force), therefore no induced current.

2. Side AB rising, value of current rising (direction = A → B).

3. Maximum number of lines of force being cut, therefore maximum value of current.

4. Side AB still rising but cutting less lines of force, therefore value of current falling (direction still A → B).

5. No lines of force being cut (movement of coil along the direction of lines of force), therefore no induced current.

6. Side CD rising, value of current rising (direction = D → C).
7. Maximum number of lines of force being cut, therefore maximum value of current.
8. Side CD still rising but cutting less lines of force, therefore value of current falling (direction still D → C).
9. No lines of force being cut (movement of coil along the direction of lines of force), therefore no induced current.

The graph we have drawn represents one complete rotation of the coil, i.e. 360°. This is known as one cycle. Because the coil continues to rotate at a constant speed we can talk about the number of cycles per second. This is known as the *frequency* of the current.

In Britain electricity is generated at 50 cycles per second (sometimes called 50 hertz, where 1 cycle per second = 1 hertz).

The shape of this graph is known as a *sine curve* (or sine wave) (although we have talked about current it should be realised that the e.m.f. produced will follow the same shaped curve).

PEAK AND EFFECTIVE VALUES OF ALTERNATING CURRENT

When we considered direct current it was easy to measure the value of potential difference and current since they do not vary with time, i.e. they had constant values. This is however not the case with alternating current.

Peak values. These are the maximum values of voltage or current (either positive or negative). They are represented by points 3 and 7 in Fig. 10/2.

It is useful to know these (a) if we need to calculate the minimum wavelength of x-rays produced by an x-ray tube operating at a certain kilovoltage (Chapter 15), and (b) for determining the insulation requirements around a conductor.

However, peak values have few other uses.

Average values. These are even less useful than peak values, the average value being zero since positive and negative values on the graph are equal.

Effective values. This is an 'average' value which will represent the effect of the current, for example in terms of its heating effect (power). This is known as its *effective value* and it is defined as follows: 'the effective value of an alternating current is that value of direct current which will produce the same heating effect'. For example, an alternating current whose peak value is 14 amps will have the same heating effect as a direct current of 10 amps. So we say that the effective value of the alternating current is 10 amps.

RELATIONSHIP BETWEEN PEAK AND EFFECTIVE VALUES

Peak value = effective value × $\sqrt{2}$ (where $\sqrt{2}$ is approx. 1.4).

$$\text{Effective} = \frac{\text{peak}}{\sqrt{2}} \text{ (or peak} \times 0.7\text{).}$$

Effective values are often known as RMS values (this stands for Root-Mean-Square and it represents the mathematical way in which the effective value is derived).

Here are two simple calculations:

1. The mains voltage is 240 volts RMS. Calculate its peak value.

$$\begin{aligned}\text{Peak} &= \text{RMS} \times 1.4 \\ &= 240 \times 1.4 \\ &= 336 \text{ volts peak.}\end{aligned}$$

2. An alternating current whose peak value is 8 amps flows through a resistance of 5 ohms. Calculate the power.

$$\begin{aligned}\text{Power} &= I^2R \\ &= \frac{8}{\sqrt{2}} \times \frac{8}{\sqrt{2}} \times 5 \\ &= \frac{64}{2} \times 5 \\ &= 160 \text{ watts}\end{aligned}$$

$$\text{(remember } \sqrt{2} \times \sqrt{2} = 2\text{).}$$

Note that it is necessary to convert peak values of current to RMS since peak values represent the value of current at one moment in time only, whereas RMS represent the effective value of the current.

TRANSFORMERS

In Chapter 9 we showed how electrical energy could be transferred from one circuit to another by means of a link of changing magnetic flux; see Fig. 9/5 and you will remember that an e.m.f. was induced across the secondary coil when the switch was opened or closed, i.e. when the magnetic field was changing.

Let us now consider the implication of replacing the battery and switch with a source of alternating current. Since the current is continually changing in both value and direction, the magnetic field associated with this current will also be changing. Therefore an alternating e.m.f. will be induced across the secondary coil.

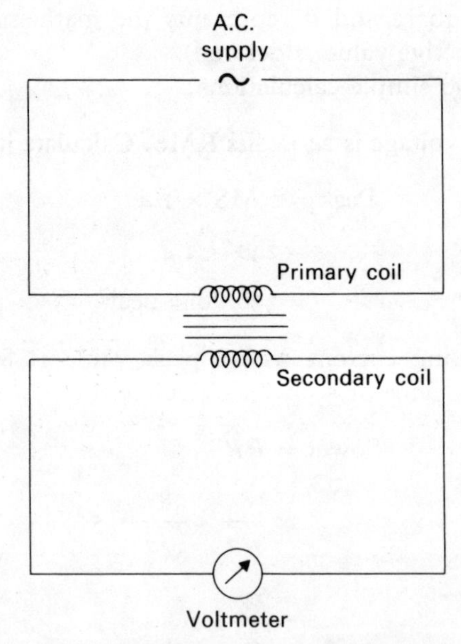

Fig. 10/3. Mutual induction using A.C. supply.

CONSTRUCTION OF A SIMPLE TRANSFORMER

Fig. 10/4a shows the construction of a simple transformer. The primary and secondary winding are wound on a soft iron core, this is to assist in transferring the magnetic field from the primary to the secondary. Each of the coils or turns of wire is coated with varnish to insulate it from the others and from the iron core.

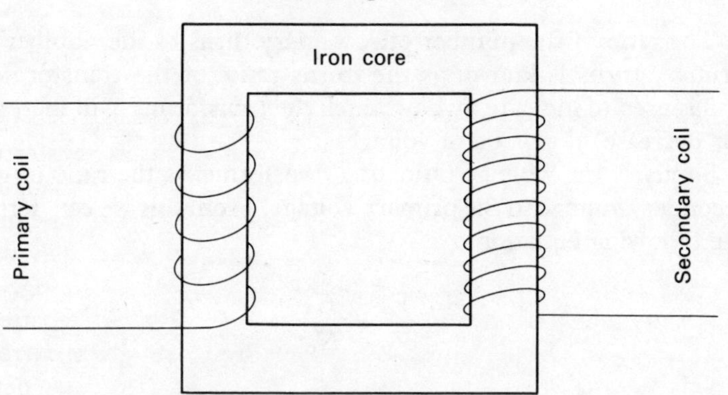

Fig. 10/4. (a) Simple transformer.

HOW A TRANSFORMER CHANGES THE VALUE OF VOLTAGE

Let us consider a simple transformer like the one in Fig. 10/4a having 5 turns on the primary and 10 turns on the secondary.

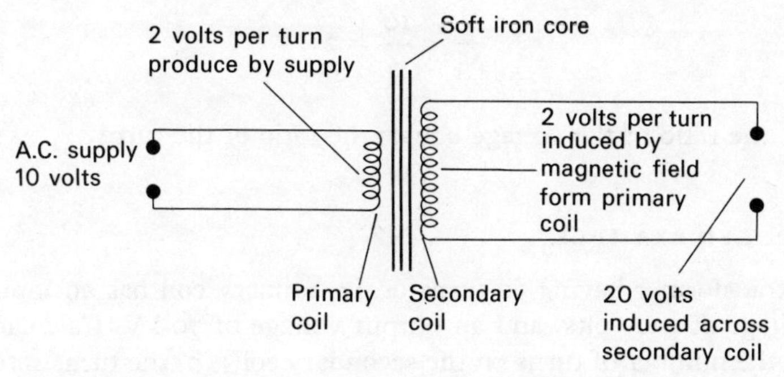

Fig. 10/4. (b) How a transformer can change the value of voltage.

If 10 volts are applied across the primary, it is reasonable to suppose that there will be 2 volts across *each* turn (the PD of 10 volts is divided equally across each of the 5 turns). This will produce a magnetic field whose strength will induce 2 volts on each turn of wire that it cuts. So if there are 10 turns on the secondary it will produce 20 volts.

Thus we have 'transformed' 10 volts into 20 volts. We can do the same for any value of voltage providing we select the corresponding number of turns on the primary and secondary coils.

A transformer which increases voltage as in the above example is known as a *step-up* transformer.

The ratio of the number of secondary turns to the number of primary turns is known as the 'turns ratio' of the transformer, and is used to indicate by how much the transformer will increase (or decrease) the value of voltage.

Similarly the 'voltage ratio' of a transformer is the ratio of the secondary voltage to the primary voltage. From this we can derive the following equation:

$$\frac{V_s}{V_p} = \frac{N_s}{N_p}$$

where V_s = secondary voltage; V_p = primary voltage; N_s = number of turns on secondary coil; N_p = number of turns on primary coil. Applying this to our previous example:

$$\frac{V_s}{V_p} = \frac{20}{10} = 2,$$

$$\frac{N_s}{N_p} = \frac{10}{5} = 2$$

i.e. the ratio of the voltage equals the ratio of the turns.

WORKED EXAMPLE

A transformer having 60 turns on its primary coil has an input voltage of 240 volts, and an output voltage of 70 kV. Calculate (a) the number of turns on the secondary coil; (b) the turns ratio of the transformer.

(a) $\frac{V_s}{V_p} = \frac{N_s}{N_p}$

$$\frac{70\,000}{240} = \frac{N_s}{60}$$

$$240 \times N_s = 60 \times 70\,000$$

$$N_s = 30\,000.$$

(b) Turns ratio $= 30\,000 : 60$

$$= 500 : 1.$$

(Note this calculation assumes that the quoted voltages are either both peak or both RMS. If this is not the case a conversion is necessary using peak = RMS × $\sqrt{2}$.)

In our first example we transformed 10 volts into 20 volts. This suggests that we are getting something for nothing! This however cannot be true (Law of Conservation of Energy; see Chapter 1). What we can say, however, is:

$$\text{Power output} = \text{power input}$$
$$\text{(assuming 100\% efficiency)},$$

$$\text{and since} \quad \text{Power} = V \times I$$

$$\text{Therefore} \quad V_s \times I_s = V_p \times I_p$$

(i.e. power in secondary, $V_s I_s$ = power in primary, $V_p I_p$).

It follows, therefore, that if V_s is increased I_s must be decreased. Looking again at our example:
If we extract 10 watts of power from the secondary, the primary will draw 10 watts from the supply, i.e. if we require ½ amp at 20 volts from the secondary the primary would draw 1 amp at 10 volts, thus what we gain in voltage we lose in current, but since we need high voltages and only small currents to produce x-rays this situation is quite satisfactory.

We have said that transformers which increase voltage are called step-up transformers. Such a transformer has its turns ratio written, for example, as 2 : 1. This means that it is a step-up transformer with twice as many turns on the secondary as on the primary.

STEP-DOWN TRANSFORMERS

This type of transformer reduces the voltage. For example, a step-down transformer with 25 turns on the primary and 5 turns on the secondary would step the voltage down 5 times and would have its turns ratio written as 1 : 5.

EFFECTS OF LOAD ON A TRANSFORMER

The load on a transformer is the power drawn from its secondary winding. To help understand this we will use two examples:

Example (a). No-load on secondary (open circuit).

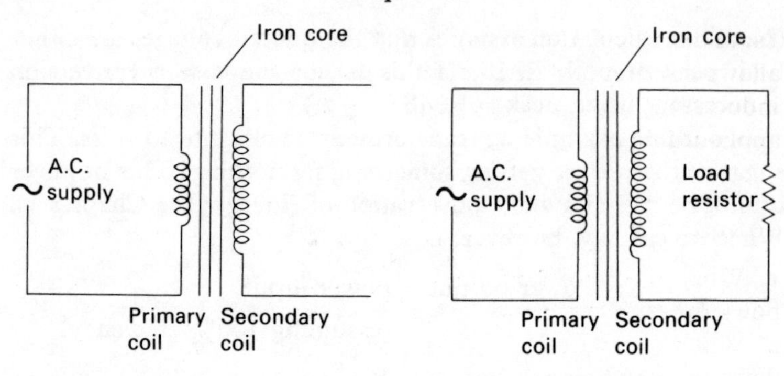

Fig. 10/5. No-load on secondary coil. Transformer on open circuit.

Fig. 10/6. Transformer on-load.

Explanation.

1. Voltage across primary produces a current through the primary.
2. This primary current produces a magnetic field around the primary.
3. This magnetic field cuts the primary and induces a voltage.
4. From Lenz's Law, this voltage is in the opposite direction to that producing it (back e.m.f.).
5. This induced voltage (back e.m.f.) thus prevents any current flowing through the primary circuit (because it is equal and opposite to the applied voltage).
6. N.B. An e.m.f. is induced across the secondary, but no current can flow because it is not a complete circuit and since no current can flow there will be no magnetic field produced by the secondary.

Example (b). Secondary on-load (complete circuit). In addition to **1** to **6** in example *(a)* the explanation continues:

7. Because the secondary is a complete circuit, the induced e.m.f. causes a current to flow.
8. This current produces a magnetic field around the secondary.
9. This magnetic field around the secondary is in the opposite direction to that around the primary, and therefore tends to cancel it out.

10. This reduces the back e.m.f. induced in the primary and allows a current to flow through the primary (since the voltage induced by the primary magnetic field is now less than the applied voltage).

CONCLUSIONS

1. When the secondary of a transformer is 'off-load', no current flows through the primary.
2. When the secondary of a transformer is 'on-load', current flows through the primary.
3. The greater the current drawn from the secondary, the greater the current flowing through the primary.

(These statements are true in an ideal transformer, i.e. one in which there are no losses; see below).

TRANSFORMER EFFICIENCY

So far when talking about transformers we have assumed that they are 100% efficient, i.e. power out = power in.

However, this is not the case in practice and power out is always less than power in.
Efficiency is defined by the following equation:

$$\text{Efficiency} = \frac{\text{power out}}{\text{power in}} \times 100\%.$$

TRANSFORMER LOSSES

We have just said that transformers are not 100% efficient. This implies that energy is lost somewhere in the transformer. There are three main processes responsible for these losses.

Copper losses. These are losses due to the heating effect of a current, and they are proportional to (a) the resistance of the wire ($H \propto R$); and (b) the square of the value of the current ($H \propto I^2$).

Since I will be greatest in the primary circuit in the case of a step-up transformer the windings of the primary are thicker to reduce heat production due to the resistance of the wire.

The opposite is true in the case of a step-down transformer.

Eddy currents. These are circular currents that are caused when the changing magnetic field set up in the core induces an e.m.f. within the iron. If the core is made of solid iron, its resistance will be very small, the resultant currents will therefore be large and produce a lot of heat.

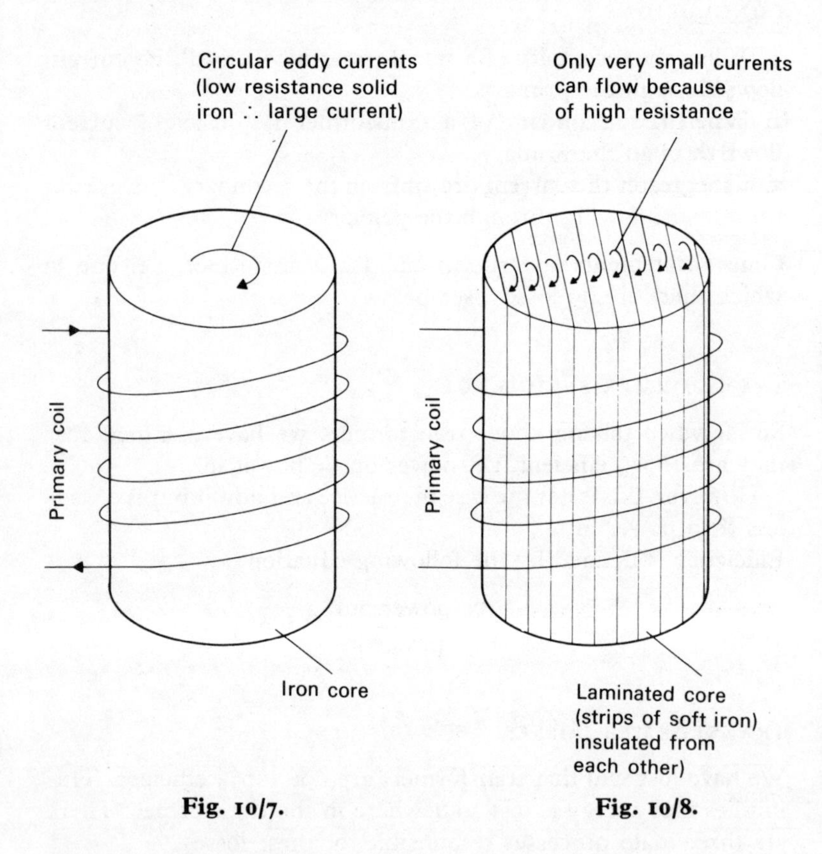

Fig. 10/7. **Fig. 10/8.**

This problem is overcome simply and effectively by making the core from many thin pieces of iron insulated from each other.

This is called a *laminated* core and it completely overcomes the problem of eddy currents. The very high resistance between laminations results in very small induced currents.

Hysteresis. When we talked about magnetism in Chapter 8 we said that all the individual domains were turned to face the same direction in a magnetised material.

If we now consider the core of a transformer we realise that as the direction of the magnetic field is constantly changing so must the direction in which the domains are facing be constantly changing. For this to occur energy is required. This energy is obtained from the primary current.

In practice these losses are minimised by choosing a material in which the domains are free to turn easily, e.g. stalloy.

To show the hysteresis effect we need a graph called a hysteresis loop (Fig. 10/9a).
In order to explain this graph we will follow it in the direction A → B → C → D → E → F → G → B (Table 10/1) and compare it with the graph of a.c. (Fig. 10/9b).

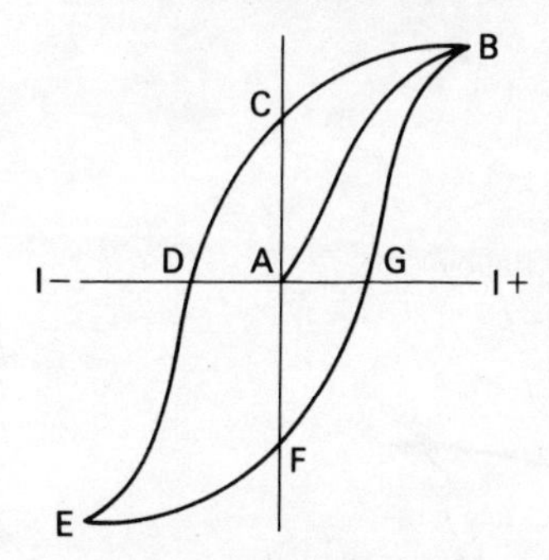

Fig. 10/9. (a) Hysteresis loop.

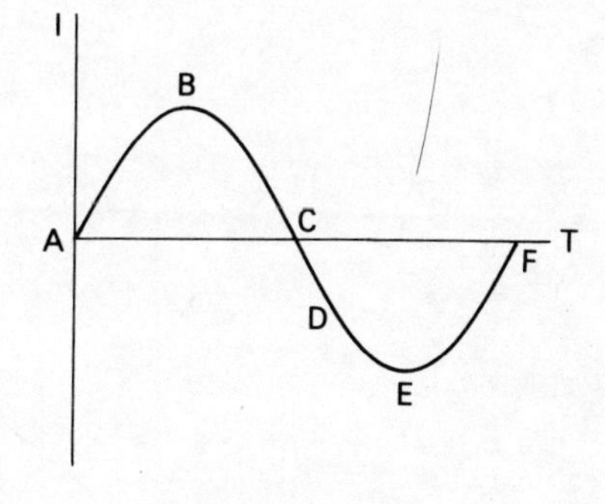

Fig. 10/9. (b) a.c. cycle.

Table 10/1.

Graph	Current	Magnetic field
A → B	Rising to max. positive	Building up forwards
B → C	Falling to zero	Collapsing
C → D	Rising negatively	Still collapsing to zero
D → E	Reaching max. negative	Reversing
E → F	Falling to zero	Collapsing
F → G	Rising positively	Still collapsing
G → B	Reaching max. positive	Building up forwards

N.B. Points to note on the graph are C and F. This is where the current is zero but there is still residual magnetism in the core.

Fig. 10/10 shows two hysteresis loops, one for soft iron and one for stalloy. The larger the area within the loop, the greater the energy consumed in turning the domains.

LOSS OF MAGNETIC FLUX

Any lines of force which are produced by the primary but do not cut the secondary will reduce the induced voltage in the secondary. To prevent the loss of lines of force, transformers are frequently wound with their secondary coil on top of the primary and on an H-shaped iron core (to correspond to the shape of the magnetic field).

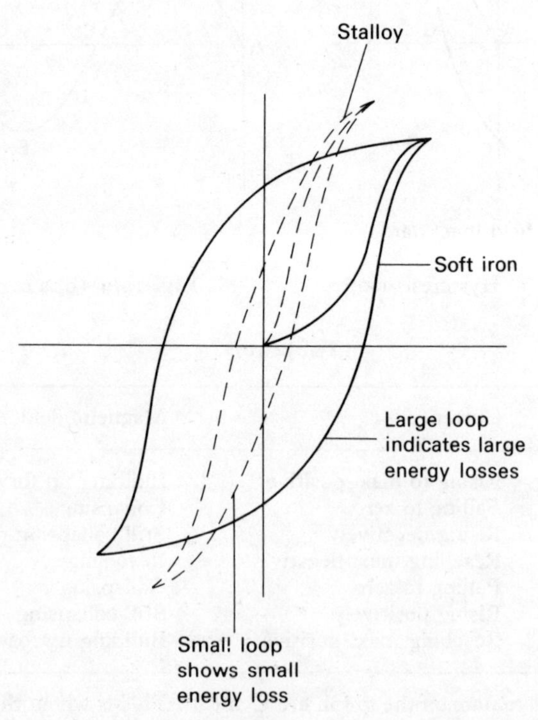

Fig. 10/10. Comparative hysteresis loops for stalloy and soft iron.

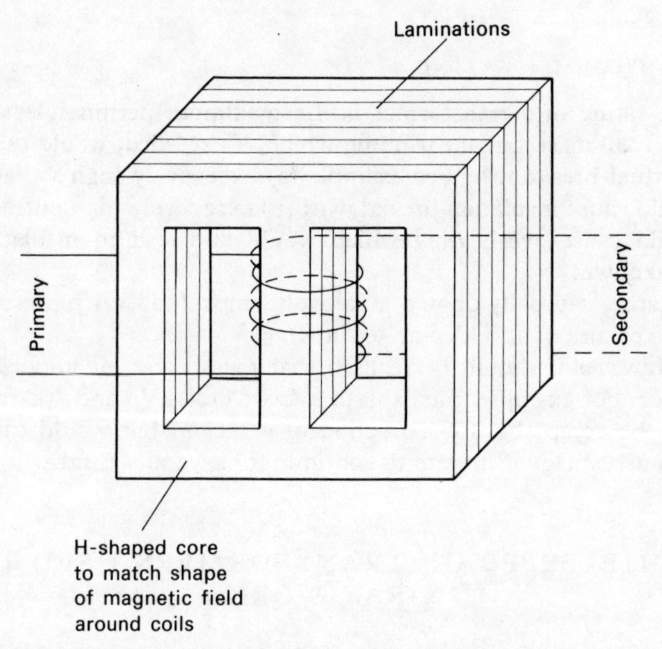

Fig. 10/11. Primary and secondary coils wound on centre leg of H-shaped core.

A transformer may have three or four secondary windings, each with a different number of turns, thus enabling more than one output voltage to be obtained.

TRANSFORMER REGULATION

When we mentioned copper losses we said that these were equal to I^2R. This means that the energy, and the voltage lost, will be greater at high currents. To assist us when we are buying a transformer, the manufacturers usually quote the regulation which is the voltage loss as a percentage, e.g. 10%. The figure is arrived at by this equation:

$$\frac{V(\text{off load}) - V(\text{on full load})}{V(\text{off load})} \times 100\%$$

Thus transformers with a low-percentage regulation are better than transformers with a high-percentage regulation.

TRANSFORMER RATING

The rating of a transformer is the maximum permissible safe output of its secondary winding which, if exceeded, would cause eventual breakdown. For example (a) excessively high voltages would cause insulation breakdown; (b) excessively high currents would cause overheating which would also lead to insulation breakdown.

Rating is usually quoted in kilovolt·amps. (50 kVA represents an exposure of 100 kVp at 500 mA).

However it should be realised that *time* is also an important factor. For example, the exposure above (100 kVp and 500 mA) may be acceptable for a fraction of a second but would cause serious damage if it were to continue for several seconds.

SOME TYPES OF TRANSFORMERS USED IN X-RAY WORK

HIGH TENSION TRANSFORMER (500 : 1 step-up)

This is used to raise the value of the mains supply to the kilovoltages required to operate the x-ray tube.

The primary winding consists of a few hundred turns (200) of thick copper wire, each turn is insulated from the others with varnish, and an insulating cylinder prevents contact with the core. The wire is thick because of the very large currents it carries during an exposure (up to 300 amps; compare that to a 1-kilowatt electric fire which draws 4 amps).

The secondary winding consists of a large number of turns (100 000) of thin copper wire coated with varnish. The wire is thin for two reasons; firstly 100 000 turns of wire occupy a large space; and secondly, the current in the secondary winding is relatively small (0.5 amp). For reasons which will be explained later, the centre of the secondary is earthed.

Because of the high voltages extra insulation is required, and this is provided in the following ways: (a) two insulating cylinders are situated between the secondary coil and the core; (b) a layer of waxed paper is included between each layer of windings on the secondary; (c) the whole transformer is immersed in oil. The oil has the following functions:

1. To improve the insulation between the windings.
2. To keep the transformer free from dust and dirt.
3. To provide a cooling agent by transfering the heat to the sides of the tank by convection.

FILAMENT TRANSFORMER (1 : 20 step-down)

The filament transformer is used to reduce the value of the mains to 12 volts and supply a current of several amps required to heat the filament of the x-ray tube. It is also found in the generator tank immersed in oil, because its secondary winding is connected to the filament of the x-ray tube, which is at a high voltage. The filament transformer also separates or isolates this high voltage on the x-ray tube from parts of the circuit which we are likely to come into contact with. For this reason it is sometimes referred to as an isolating transformer.

AUTOTRANSFORMERS

So far all the types of transformers we have talked about have had their primary and secondary windings separate, i.e. linked only by magnetism.

In the autotransformer the primary and secondary windings are joined together, with part of the winding belonging to both primary and secondary, and are wound on a single leg of laminated core.

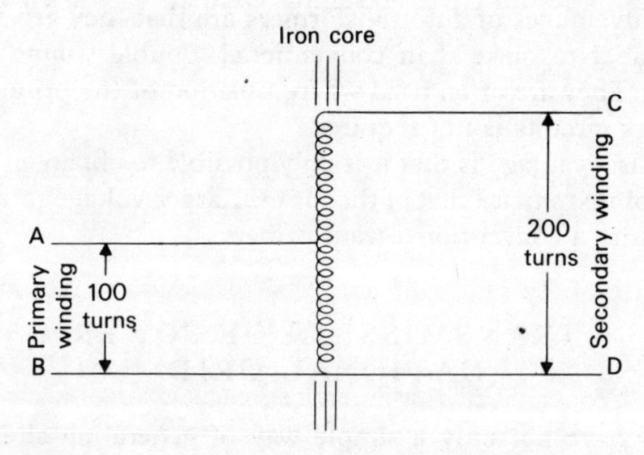

Fig. 10/12. Autotransformer (used as a step-up transformer).

The autotransformer uses the principle of self-induction.

How it works. In the above example the primary consists of 100 turns and the secondary of 200 turns. If 100 volts is applied to the primary winding, an e.m.f. of 1 volt per turn will be induced by the magnetic field.

Since this magnetic field links the 200 turns of the secondary, then 1 volt per turn will be induced across the secondary i.e. 200 volts.

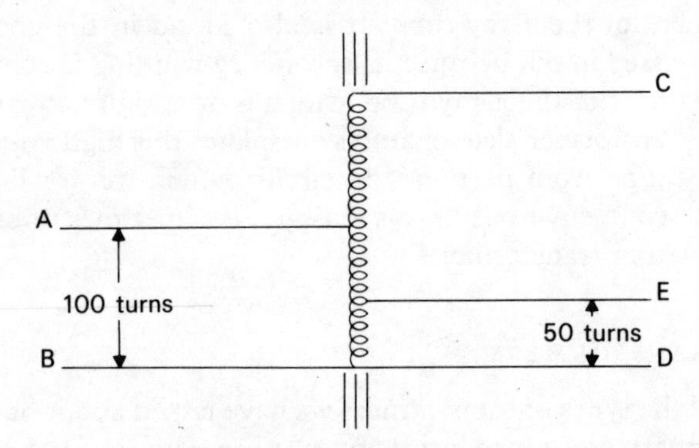

Fig. 10/13. Autotransformer (used as a step-down transformer).

By adding further tappings (connections) to the secondary coil, it becomes clear that voltages less than that across the primary can also be obtained. Again assuming 100 volts are applied to the primary, we can now obtain 200 volts from C and D, and 50 volts from D and E.

The advantages of autotransformers are that they are simpler and cheaper to make than conventional 'double wound' transformers. They are often used where isolation of the primary and secondary circuits is not required.

The disadvantage is that it is only possible to obtain an output voltage of up to twice that of the input. Larger voltage ratios than this require a conventional transformer.

TRANSMISSION OF POWER (NATIONAL GRID)

We now have not only a simple way of generating alternating current but also of changing the value of its voltage. We have the choice of transmitting this power from the generating stations to people's homes, hospitals, industries, etc. either at high voltages and low currents, or at low voltages and high currents. Let us try to discover which would be the best.

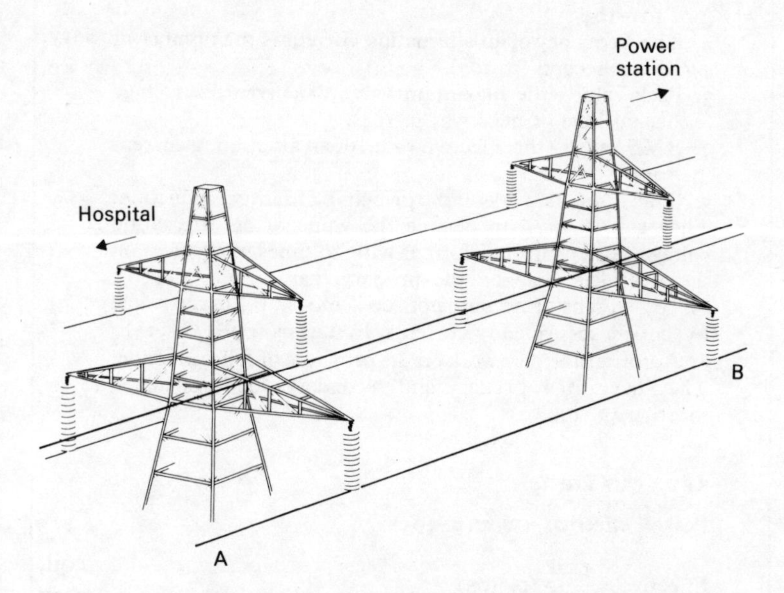

Fig. 10/14. Transmission of power.

The diagram shows e.m.f. generated at the generating station and power transmitted down the cables.

If we look at just one conductor (e.g. A → B), from Ohm's Law we know that there will be a drop in voltage between its ends, and that this drop in voltage will be proportional to the current flowing through it, i.e. the larger the current, the larger the voltage drop.

Another way of looking at this question is to remember that the heat produced is proportional to the square of the current ($H \propto I^2$). Therefore by making the current as small as possible we will reduce the power losses due to the production of heat.

For these two reasons it is therefore more economical to transmit power at *high voltages* and *low currents* (in practice power is transmitted through the national grid system at 132 000 volts).

CHAPTER SUMMARY

1. Alternating current is produced by electromagnetic induction using an alternator. A sine wave is generated (pp. 101–104).
2. The frequency of an alternating current is the number of cycles per second (p. 104).
3. Peak value—the maximum value of current or voltage (either positive or negative; p. 104).
4. RMS value—the effective value of an alternating current (p. 105).
5. Transformers work on the principle of mutual induction. They can be used to change the value of an alternating voltage. A step-up transformer with 20 times more turns on the secondary than on the primary, has a ratio of 20 : 1 (p. 107). Transformers are not 100% efficient. Energy is lost by, copper losses, eddy currents, hysteresis (pp. 111–113).
6. Autotransformers work on the principle of self-induction. They have their primary and secondary windings joined together (p. 117).

KEY EQUATIONS

Peak = effective × $\sqrt{2}$ (p. 105).

Effective = $\dfrac{\text{peak}}{\sqrt{2}}$ (p. 105).

$\dfrac{V_s}{V_p} = \dfrac{N_s}{N_p}$ (p. 108).

(ratio of secondary to primary voltage)

Chapter 11
Thermionic emission

When a current flows through a conductor, free electrons move along the wire, their direction depending on the direction of the potential difference (see Chapter 7). Now we are going to consider how we can release electrons from the wire completely.

Atoms and electrons inside a conductor are constantly vibrating. The amount of vibration depends upon the energy they possess. This energy in turn depends on the temperature of the conductor.

At room temperature these electrons can move freely within the wire because of attraction between neighbouring atoms, but are unable to leave the *surface* of the material because of the combined effect of all the other atoms tending to pull them back. As the temperature of the conductor increases, the energy of the electrons also increases until a point is reached when the electrons have sufficient energy to escape the binding forces of the atoms and then escape from the surface of the material.

This process is known as thermionic emission.

Tungsten, a material commonly used, exhibits thermionic emission at 2000°C.

The graph shows no emissions of electron until 2000°C is reached. Thereafter small increases in temperature produce large increases in electron emission. If the temperature of the material is further increased, eventually a point will be reached where some of the *atoms* may acquire enough energy to escape the surface of the material. This process is known as evaporation.

The energy needed to release electrons from the surface of a material is known as the work function.

Thermionic emission will occur more readily if the pressure of air around the material is removed, e.g. if the conductor is sealed in an evacuated glass bulb (known as a glass envelope). Let us look at what would happen under these conditions.

The most practical way of heating a conductor is to pass a current through it. As its temperature approaches 2000°C

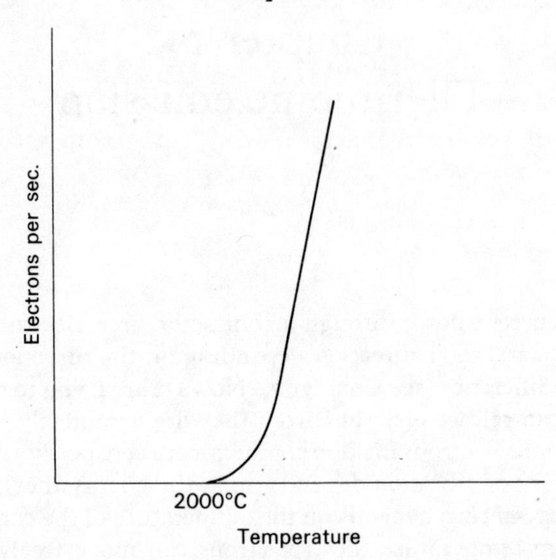

Fig. 11/1. Graph of thermionic emission against temperature (tungsten wire).

(tungsten wire) electrons will start to be emitted from the surface (a conductor is called a filament when it is used in this manner). Since all the electrons possess a negative charge they will tend to repel each other and form a cloud around the filament.

Because the conductor has emitted electrons it will tend to acquire a positive charge and will therefore attract some of the electron cloud back to itself. Thus electrons are being emitted from and returning to the surface of the filament (imagine them on elastic bands). Another effect of this negatively charged cloud of electrons is to repel further electrons which are trying to escape from the filament.

Thus when a filament reaches its emission temperature a state of equilibrium is soon reached, where the number of electrons being emitted from the filament is equal to the number of electrons returning to it.

The cloud of electrons around the filament is known as space charge, and its tendency to limit further electron emission is space charge effect.

THE DIODE VALVE

Let us add another conductor into our evacuated glass envelope. We now have a simple valve called a diode (we call these conductors electrodes). The circuit symbol is shown in Fig. 11/2.

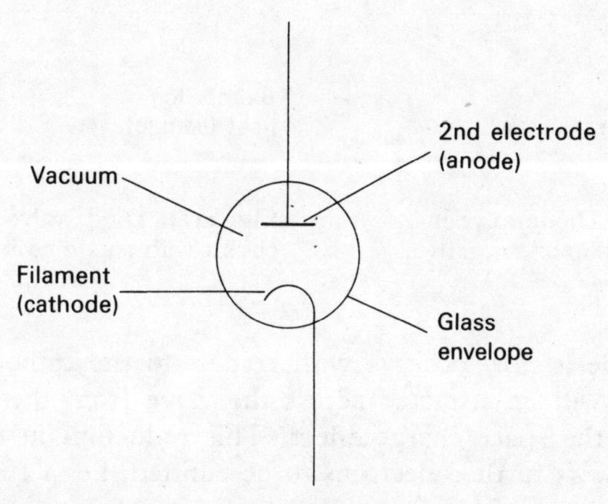

Fig. 11/2. Circuit symbol representing thermionic diode valve.

The valve comprises the following:

1. Cathode or filament (negatively charged electrode).
2. Anode (positively charged electrode).
3. Glass envelope.
4. Vacuum.

Let us connect the valve into a simple circuit.

The circuit in Fig. 11/3 comprises a battery, switch, ammeter, diode valve, high tension d.c. source and a battery to heat the filament (heating current = I_f).

As the temperature of the filament rises, the space charge effect is established. If the switch is now closed the anode will be made negative with respect to the cathode. Since no electrons can escape from the anode no current will flow through the circuit (the anode is cold and therefore will not emit electrons).

What happens if we now reverse the polarity of the battery?

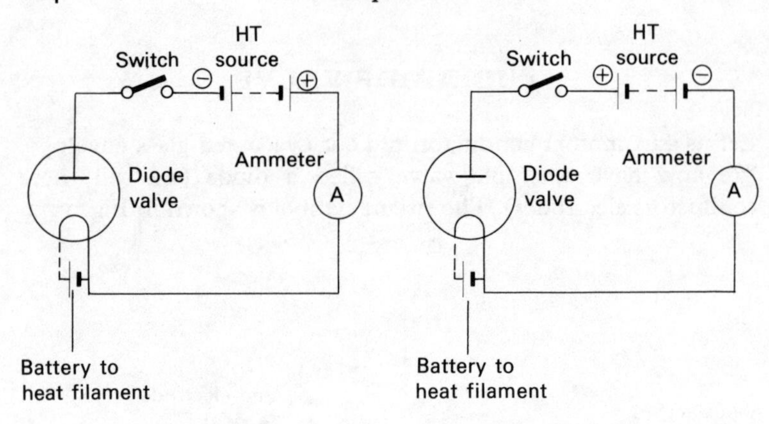

Fig. 11/3. Diode valve in circuit with anode negative.

Fig. 11/4. Diode valve in circuit with anode positive.

The anode is now positive with respect to the cathode. Thus electrons will be attracted across the valve from the filament, reducing the space charge effect. This reduction in the space charge allows further electrons to be emitted, i.e. a current (I_a) will flow through the circuit.

What we have, in effect, is a device which acts as a *conductor* of current in one direction but an *insulator* to current flowing in the opposite direction. When used in this way the valve is acting as a *rectifier*.

DIODE CHARACTERISTICS

The diode valve has many uses in x-ray physics—not the least of which is to produce x-rays. An x-ray tube is a modified diode valve. It is, therefore, important to study the diode valve in some depth.

Suppose that the rheostat is adjusted so that the filament emits a moderate number of electrons per second and we start with the anode potential (V_a) at zero and gradually increase it.

1. Initially no current (I_a) will flow since there is no PD across the valve to attract the space charge away.

2. As the anode potential rises, the anode current also rises, slowly at first, then more rapidly until a point is reached where

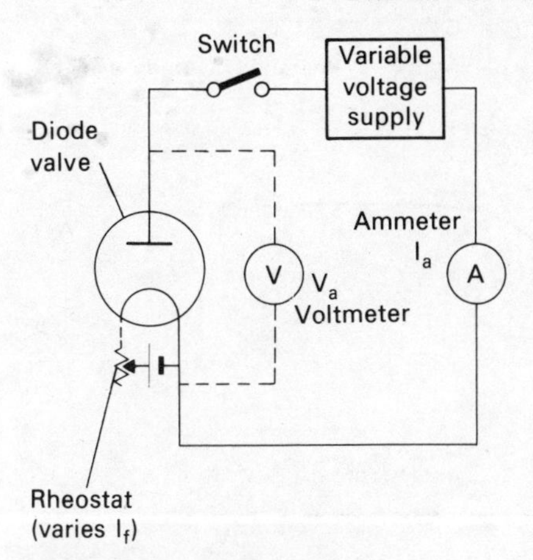

Fig. 11/5. Experiment to demonstrate diode characteristics. I_f = Filament heating current; V_a = Anode voltage (voltage across valve); I_a = Anode current (current through valve).

Fig. 11/6. Graph of anode voltage against anode current (fixed filament temperature).

further increases in V_a do not produce any increase in I_a. This is because all available electrons emitted from the filament are being attracted across to the anode.

This situation is shown graphically in Fig. 11/6.

1. Space charge limited. This is where some of the electrons from the space charge are drawn across to the anode. In this region increases in V_a *produce* increases in I_a.

2. Saturation. All electrons which are emitted from the filament are drawn across the valve.

In this region increases in V_a *do not produce* increases in I_a. This is the condition under which x-ray tubes operate so that increases in kV *do not produce* increases in mA.

In the saturated portion of the graph increases in I_a can only be obtained by increasing the filament temperature. This is done by raising the filament heating current (I_f).

Look at what happens to the graph when the filament temperature is increased (Fig. 11/7).

We see that at temperature T_2 (increased temperature from T_1) once again saturation is reached, but saturation current is higher.

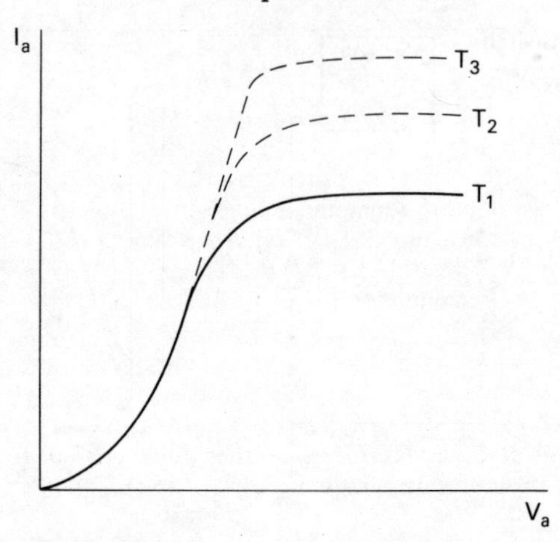

Fig. 11/7. Effect of varying filament temperature.

Raising the temperature again to T_3 gives a further rise in saturation current. This important graph is known as the characteristic curve for a diode valve.

USES OF DIODES

We have already mentioned that a diode valve is used for rectification. In Chapter 13 we will look at this more closely, with particular reference to the construction of the x-ray tube.

TYPES OF VALVES

Other types of valve such as the vacuum triode valve with three electrodes and various forms of gas-filled valves are beyond the scope of this book.

CHAPTER SUMMARY

1. Thermionic emission is the release of electrons from the surface of a conductor by the action of heat (p. 121).
2. Space charge is the cloud of electrons around the filament in a valve (p. 122).
3. The space charge limited condition is when the anode current is limited by the space charge effect (p. 125).
4. Saturation is the condition in which all available electrons are being drawn across the valve. Further increases in anode voltage will not produce increases in anode current (p.125).
5. Diode valves are rectifiers, i.e. they allow current to flow through them in one direction only (p. 124).

Chapter 12
Stationary anode x-ray tube

In Chapter 11 we said that an x-ray tube was a special type of diode valve. To meet the requirements of producing x-rays, some modifications have been made, particularly to the size and shape of the anode and cathode.

To produce x-rays we need to decelerate fast-moving electrons very quickly. The electrons are accelerated across to the positive anode which they hit at high speed. As they are brought to rest by the collision some of their energy is converted into x-rays.

More than 99% of the energy released by these electrons appears as heat; it is clear that heat is going to be a major consideration in the design of the anode. Another problem, particularly in diagnostic work, is the size of the source of x-rays. For example, imagine two sources of x-rays: (a) a small point source; and (b) a larger finite source.

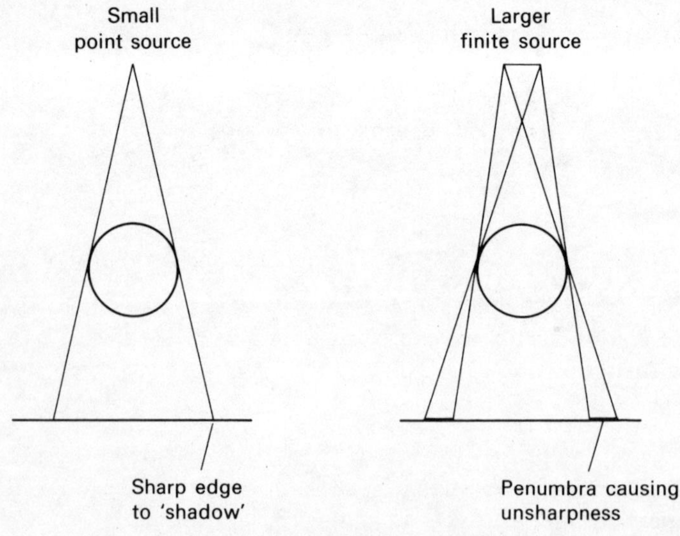

Fig. 12/1. Effect of small and large focal spot sizes.

In example (a) a sharp shadow is cast. Contrast this with example (b) where the shadow is no longer sharp; the edges of the image are seen to be spread out, not clearly defined as in (a). This unsharpness is known as *penumbra effect.*

To overcome this problem we ideally need a point source of x-rays, but this presents us with two additional problems; firstly, if all the electrons strike one point on the anode, the heat produced at this point would cause the anode to melt; and secondly, how can we get all the electrons to arrive at one point, remembering that they are all negatively charged, and will therefore tend to repel each other as they travel from the cathode to the anode, and therefore spread out.

To see how these problems are overcome, let us examine the construction of the stationary anode x-ray tube.

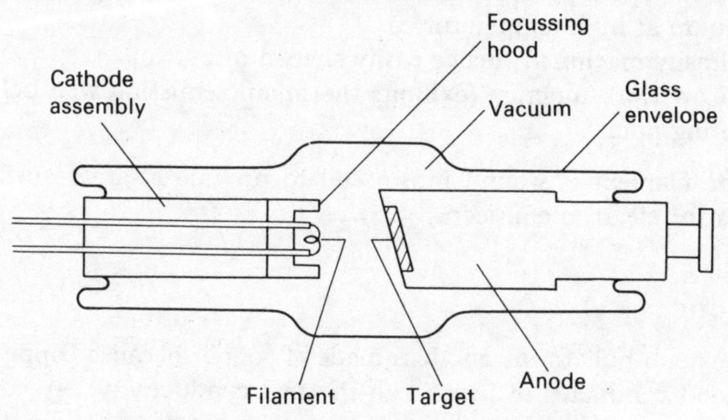

Fig. 12/2. Stationary anode x-ray tube.

CONSTRUCTION

VACUUM

All the air is excluded from within the glass envelope, so that there is nothing for the electrons to collide with, as this would slow them down.

GLASS ENVELOPE

This is made of boro-silicate glass (pyrex) to withstand the high temperatures. It also acts as an electrical insulator between the negative cathode and the positive anode. It contains the vacuum.

CATHODE

The cathode consists of a coil of wire enclosed in a focussing hood. The function of the focussing hood is to produce a parallel beam of electrons crossing to the anode. This is achieved by connecting one of the wires supplying the filament to the focussing hood, thus raising the potential of the hood to that of the filament, which is at a high negative potential. The edges of the hood produce an electric field which is responsible for this focussing effect.

Tungsten is chosen for the filament material for the following reasons:

1. High melting point.
2. Low vapour pressure (less likely to release tungsten gas into the vacuum at high temperatures).
3. Easily machined; hence easily shaped into a coil.
4. Low work function (exhibits thermionic emission well below melting point).

(The filament is wound into a coil to provide a larger surface area for electron emission.)

ANODE

The main bulk of the anode is made of copper because copper is a good conductor of heat (high thermal conductivity). It has a large mass to allow it to absorb more heat, and transmit it quicker. Other characteristics of copper however make it a bad choice for the source of x-rays (particularly its low melting point and its low atomic number; see Chapter 15). A tungsten block is therefore inset into the surface of the anode and the electrons are made to strike this. This is known as the target.

The reasons for the choice of tungsten for target material are as follows:

1. High atomic number (efficient at producing x-rays).
2. Reasonable thermal conductivity (helps to conduct heat away from the target to the mass of the anode).
3. High melting point.
4. Low vapour pressure.
5. Easily machined.

You will notice that in Fig. 12/3 the target is at an angle of 17° to the electron beam. The reasons for this are most important. If we look back to our ideal situation, namely (a) a large area over which to spread the heat; and (b) a small area to give a point source of x-rays, some compromise must be reached.

PRINCIPLE OF LINE FOCUS

XY represents a parallel beam of electrons 1 mm wide × 3 mm high arriving at the target. It hits a rectangle 1 mm × 3 mm. This *real* area is represented by ABCD.

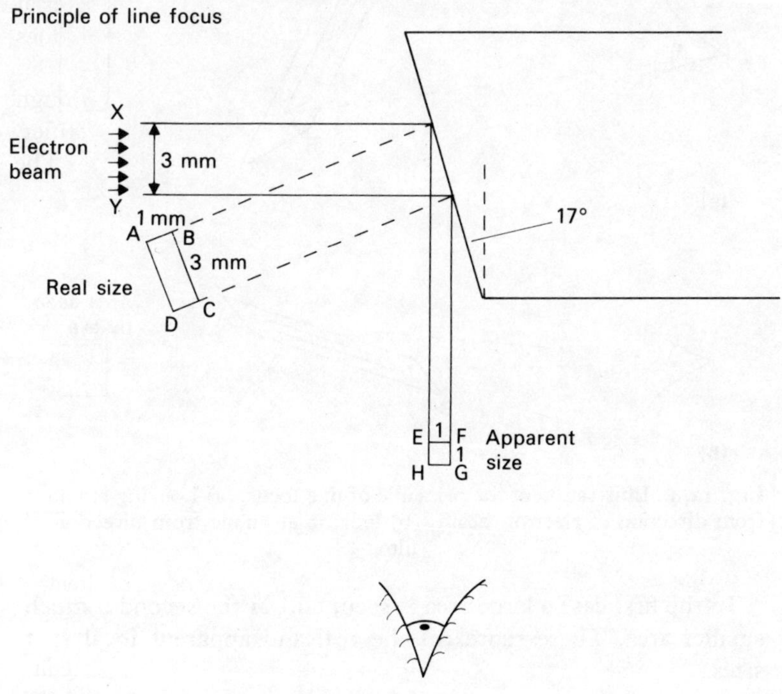

Fig. 12/3. Principle of line focus.

If we now look up at the source of x-rays from the viewpoint of the patient, the area appears to be much smaller than the real area, because of the effect of the 17° angle. In fact the area producing x-rays appears to be 1 mm × 1 mm. This is known as the apparent (or effective) focal spot size.

This effect is known as the *principle of line focus*.

You can demonstrate this principle for yourself by drawing a rectangle say 3 cm × 1 cm on a piece of card and colouring the area. This represents the area of the target that the electrons strike. Now look first at the card from the direction that the electrons would approach, holding the card at an angle of roughly 17°, and then turn the card and look at it from the position of the patient.

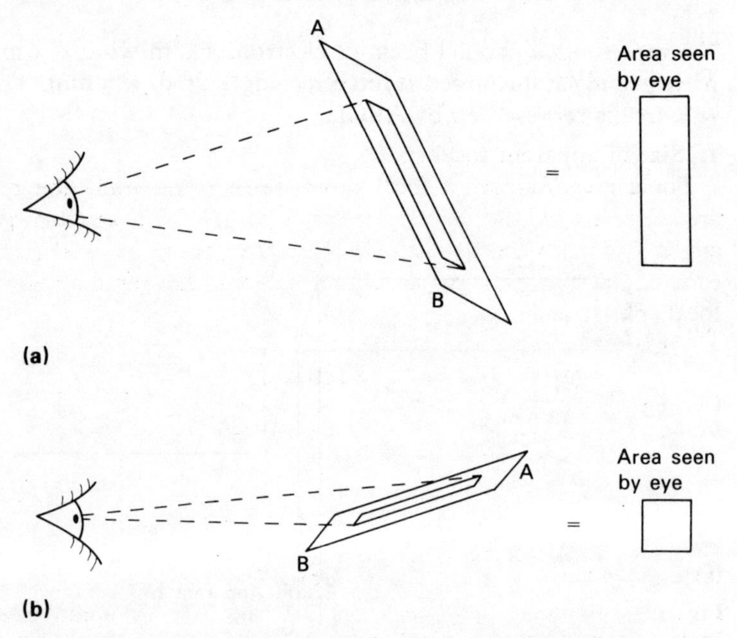

Fig. 12/4. Illustration of the principle of line focus. (a) Looking at anode from direction of electron beam; (b) looking at anode from direction of film.

In the first case a large area is seen, and in the second a much smaller area. These represent the real and apparent focal spot sizes.

EFFECTS OF CHOOSING SMALL APPARENT FOCAL SPOT SIZES

From what we said earlier about unsharpness being caused by the penumbra effect, it would seem sensible to use the smallest possible apparent focal spot size for all examinations. However, it should be realised that choosing a very small apparent focal spot size means that the real area will also be reduced in size (or the target angle must be very small; see below).

To prevent the target melting, only small values of current (e.g. 50–100 mA) can be used to operate the x-ray tube when using very small focal spot sizes (e.g. 0.3, 0.6, 1.0 mm).

If we wish to use high mA values (e.g. 400–1000 mA) larger focal spot sizes (e.g. 2.0 mm) must be used.

EFFECTS OF CHOOSING SMALL TARGET ANGLES

There are two important effects associated with the choice of target angle.

1. Size of apparent focal spot.

For a given apparent focal spot size (e.g. 1.2 mm) the real area covered by the electron beam is larger for a small target angle. Compare the height of the electron beams A_1 and A_2 in each of the above diagrams, each producing the same apparent focal spot size.

2. Area covered by x-ray beam.

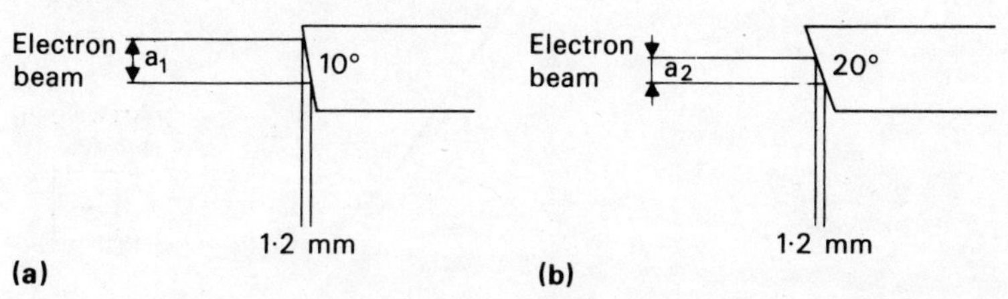

Fig. 12/5. Effect of target angle on focal spot size. (a) Small target angle—1.2 mm apparent focal spot; (b) large target angle—1.2 mm apparent focal spot.

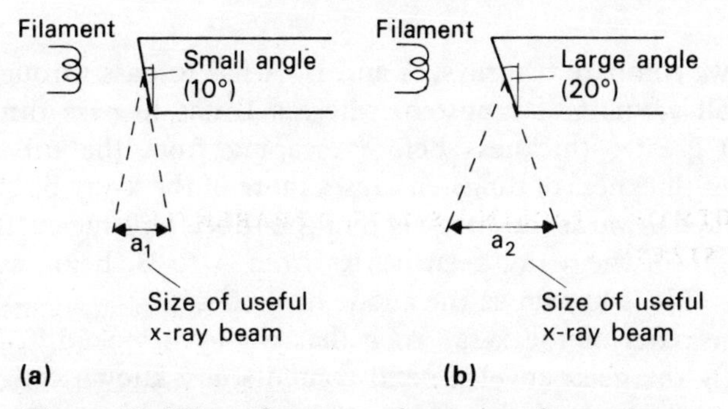

Fig. 12/6. Effect of target angle on size of x-ray beam.

As can be seen from the above diagrams the extent of the x-ray beam is limited by the angle of the face of the target (x-rays will not pass easily through tungsten or copper). This means that it is not possible to cover large x-ray films at short focus-film distances (FFDs) with a 10° target angle. (It is for this reason that radiotherapy tubes use target angles of 35° in order to cover large areas of the patient at short FFDs.)

Anode heel effect. Close examination of the x-rays emitted from the target shows that because they are produced below its surface they have to pass through some tungsten before they can escape from the tube.

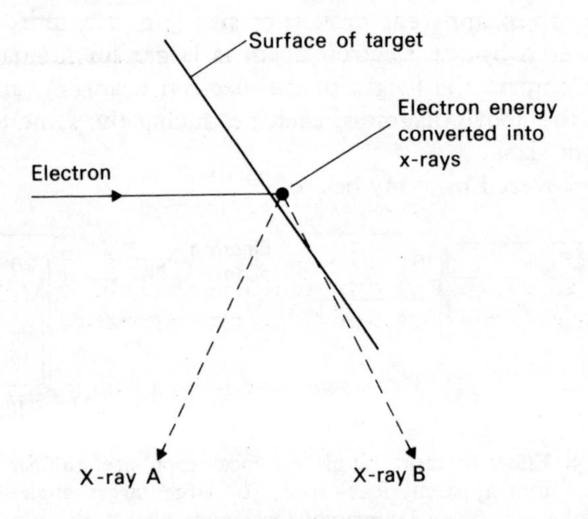

Fig. 12/7. Anode heel effect.

If we compare two rays, A and B, A has to pass through only a small amount of tungsten, whereas B has to pass through a much greater thickness before escaping from the tube. This greater thickness of tungsten causes more of the x-ray beam to be absorbed towards the anode end of the tube. This means that the strength of the x-ray beam varies from A to B, being stronger at A. This is known as the anode heel effect.

The parts of the x-ray tube that we have looked at so far, namely the glass envelope and contents, are known collectively as the tube *insert*.

X-RAY TUBE SHIELD

The tube insert is enclosed in a metal 'box', known as the shield.

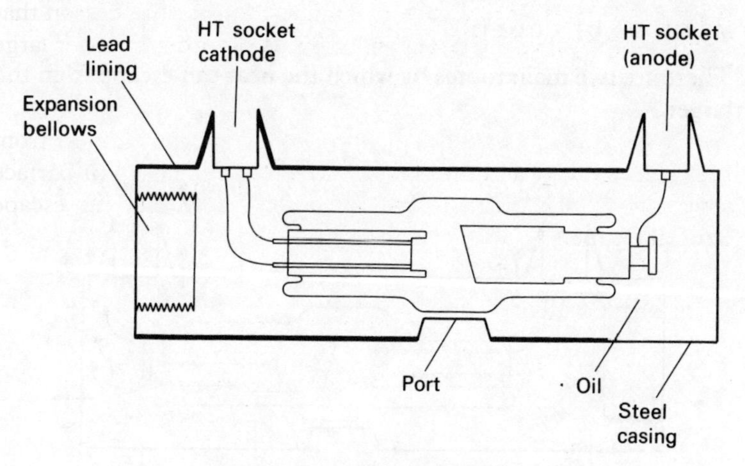

Fig. 12/8. X-ray tube shield.

It is designed to perform three functions: (a) radiation protection; (b) electrical protection; and (c) thermal protection.

Radiation protection. The steel casing is lined with lead to prevent radiation emerging in all directions. A perspex port is provided to allow the x-ray beam to escape in the useful direction only.

Electrical protection. To prevent patients and staff coming into contact with high voltages, the shield is earthed. Where the high tension cables enter the shield, insulated sockets are used.

The shield is filled with pure mineral oil which acts as an electrical insulator, preventing sparking across the insert, or from the insert to the shield.

Thermal protection. Because the anode reaches very high temperatures, the oil also acts as a cooling medium (see below).

To enable the oil to expand when it gets hot, expansion bellows are included in the shield. These allow the oil to expand and contract as its temperature changes, without letting any air enter the insert. In some types of equipment a switch is included so that when the bellows are fully compressed the switch prevents further exposures being made until the oil has had time to cool down.

METHODS OF COOLING

There are two main routes by which the heat can escape from the target.

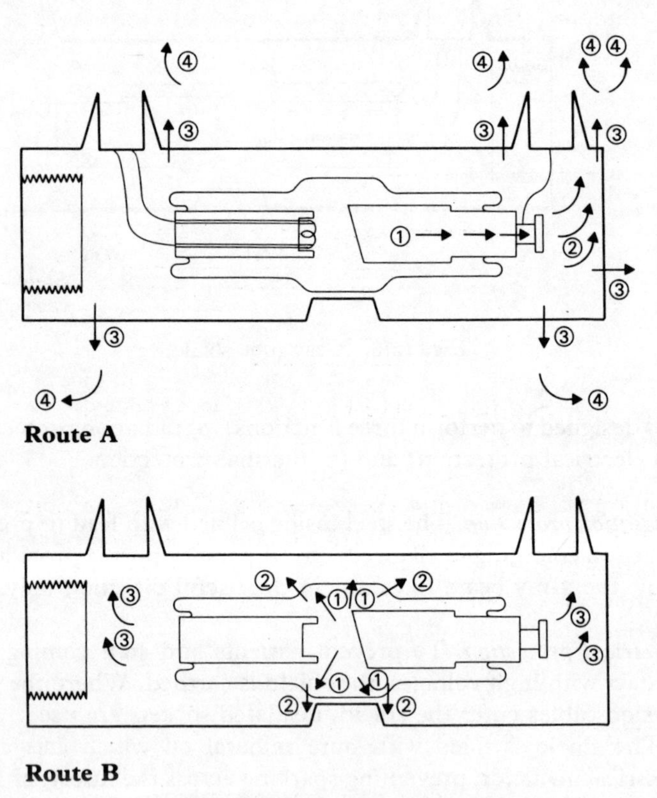

Fig. 12/9. Cooling routes.

Route A.

1. Heat is conducted down the anode stem and is transferred to the oil.

2. Convection currents in the oil transfer the heat to the shield.

3. Heat is conducted through the shield.

4. Convection currents in the air remove the heat from the shield.

Route B.

1. Heat is radiated through the vacuum to the glass envelope.
2. Heat is conducted through the glass to the oil.
Steps **3**, **4** and **5** follow the same route as steps **2**, **3** and **4** of Route A. Route A is the most important in stationary anode tubes.

In some radiotherapy tubes however, the heat produced at the anode is too great and more positive means of cooling have to be employed. One such method is shown in Fig. 12/10.

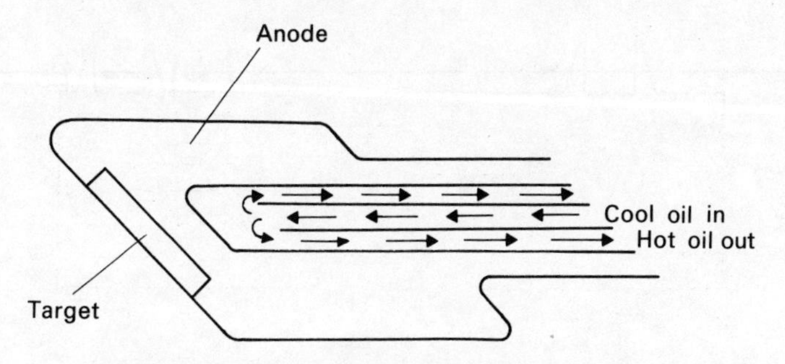

Fig. 12/10. Oil cooling in radiotherapy tubes.

Cool oil is pumped into the anode block where it absorbs heat from the anode, cooling it down. The heated oil is then cooled and returned to the anode.

Anodes cooled in this way are specially designed, having hollow interiors.

CHAPTER SUMMARY

1. Compromise is required in the design of a stationary anode x-ray tube because (a) a small focal spot is required for sharpness of the image; (b) a large focal spot is required for adequate heat dissipation (p. 131).
2. Line focus principle—the target face of a diagnostic x-ray tube, is set at an angle of about 17° so that the real area that the electrons strike is relatively large but the area of x-ray emission appears small (p. 131).

3. Small target angles (a) produce smaller apparent focal spot sizes; (b) reduce the area that the x-ray beam covers; and (c) provide a larger real focal area for any given apparent focal spot size (p. 133).
4. The anode heel effect produces falling-off in strength of the x-ray beam towards the anode end of the tube (p. 134).
5. The x-ray tube insert is enclosed in a shield designed to give (a) radiation protection; (b) electrical protection; and (c) thermal protection (p. 135).
6. Conduction, convection and radiation all play a part in the cooling of the x-ray tube (p. 136).

Chapter 13
Rectification

In Chapter 11 we said that the diode valve acts as a rectifier, i.e. a device which will pass current in one direction only. This property is particularly useful when we want to operate from the a.c. mains, equipment which works more efficiently on direct current.

In Fig. 13/1 the diode valve will conduct only when the cathode is at a negative potential with respect to the anode, because only the heated cathode is a source of electrons. Therefore current will pass through the component X in the circuit only in the direction of the arrow.

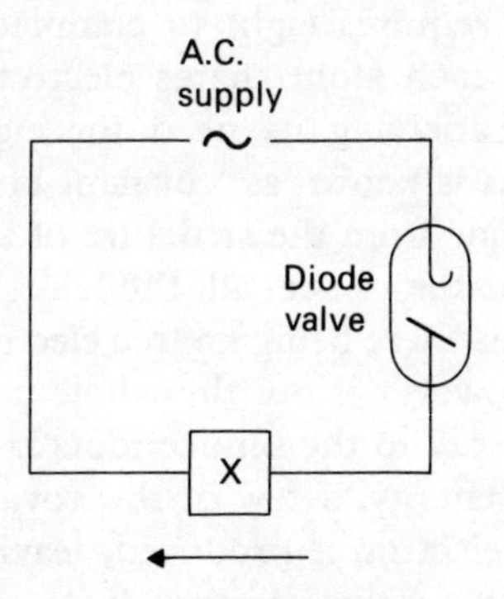

Fig. 13/1. Use of a rectifier in a circuit. For example, imagine X is a device which only functions correctly when electrons flow through it in the direction of the arrow.

Until 20 years ago, diode valves were used in most x-ray sets to provide rectification. In modern sets these have now been replaced by solid-state rectifiers, using semiconductor materials which exhibit the same characteristics as diode valves, as well as offering many advantages such as size and robustness. (The term 'solid state' refers to its solid structure in comparison with vacuum valves.)

To understand how solid-state rectifiers work it is first necessary to look at semiconductors in more detail.

PRINCIPLE OF SEMICONDUCTORS

Semiconductors, as the name suggests, have resistance values somewhere between conductors and insulators. Examples of common semiconductors are silicon, selenium, and germanium.

To understand the principles of semiconductors it is first necessary to look at their atomic structure, in particular their outermost electron shells, or valence shells.

In semiconductors we find that neighbouring atoms share their valence electrons. For example silicon has four valence electrons of its own, but it requires eight to complete its valence shell. To overcome this each atom shares electrons with four of its neighbours, thus satisfying its need for eight electrons. This sharing of electrons is known as 'covalent bonding'.

You might imagine from the structure of silicon that it would be a perfect insulator, since all the electrons are linked to neighbouring atoms, there being no free electrons to permit a flow of current. This however is not the whole truth as we will see.

If we supply energy to the semiconductor material; e.g. heat, light, or electrical energy, a few of the covalent bonds become broken and a free electron is produced, leaving a 'hole' behind. The hole is filled by another electron from a neighbouring atom which, in turn, leaves another hole and so on.

Thus we can imagine electrons and holes wandering about within the material continuously. As electrons are negative charges, we can consider holes as being positive charges. This will be a useful concept later on.

Imagine now the effect of applying a potential difference to the semiconductor.

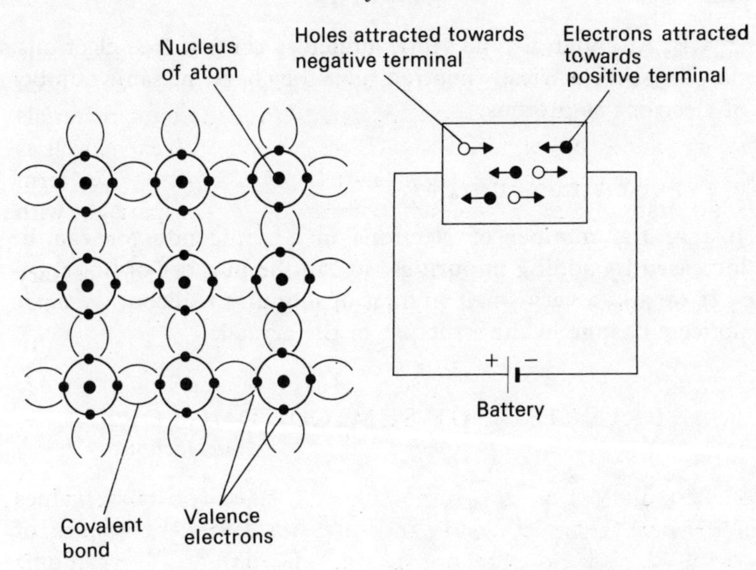

Fig. 13/2. Silicon crystal.

Fig. 13/3. Conduction in semiconductors.

As expected, the electrons move towards the positive potential and the holes towards the negative, i.e. we have a small flow of current. In a pure semiconductor there is one free electron and one hole per 1000 million atoms.

N-TYPE SEMICONDUCTORS

The number of free electrons (conductivity) of a semiconductor can be altered by adding atoms of other materials.

If we add a very small amount of antimony to silicon (e.g. 1 part per million) we find that a change in the structure of the crystal occurs.

Silicon has four valence electrons whereas antimony has five. Four of the valence electrons of the antimony atom form covalent bonds with the silicon atoms, leaving the fifth valence atom 'free' We have therefore increased the conductivity of the silicon (i.e. decreased its resistance) by the introduction of antimony.

By adding impurities which have five valence electrons (such as antimony) we produce an n-type semiconductor.

N.B. Although n-type semiconductors contain free electrons, they are still electrically neutral, since they have the same number of electrons as protons.

P-TYPE SEMICONDUCTORS

Just as the number of electrons in a semiconductor can be increased by adding impurities, so can the number of holes.

If we add a very small amount of indium to silicon, we again notice a change in the structure of the crystal.

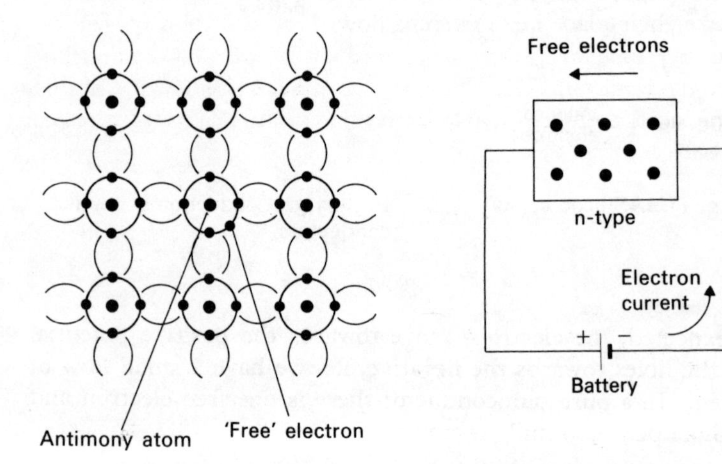

Fig. 13/4. N-type semiconductor material.

Fig. 13/5. Conduction in n-type semiconductors.

Silicon atoms have four valence electrons and indium atoms have three. The three valence electrons of indium form covalent bonds, leaving a hole where the fourth should be. We have therefore increased the conductivity of the silicon by the introduction of a hole, which is free to move and combine with other electrons.

By adding impurities such as indium whose atoms have three valence electrons we produce a p-type semiconductor.

N.B. Once again this material is still electrically neutral. Remember also that electrons and holes are wandering around continuously in the materials.

CONDUCTION IN P- AND N-TYPE SEMICONDUCTORS

Let us look at what happens when a battery is connected across an n-type semiconductor (Fig. 13/6). Free electrons flow out of the semiconductor to the positive terminal of the battery and electrons from the negative terminal flow into the semiconductor to take their place i.e. a current flows.

In the case of p-type semiconductors, current again flows through the circuit. This time we can imagine holes flowing out of the semiconductor material towards the negative terminal of the battery.

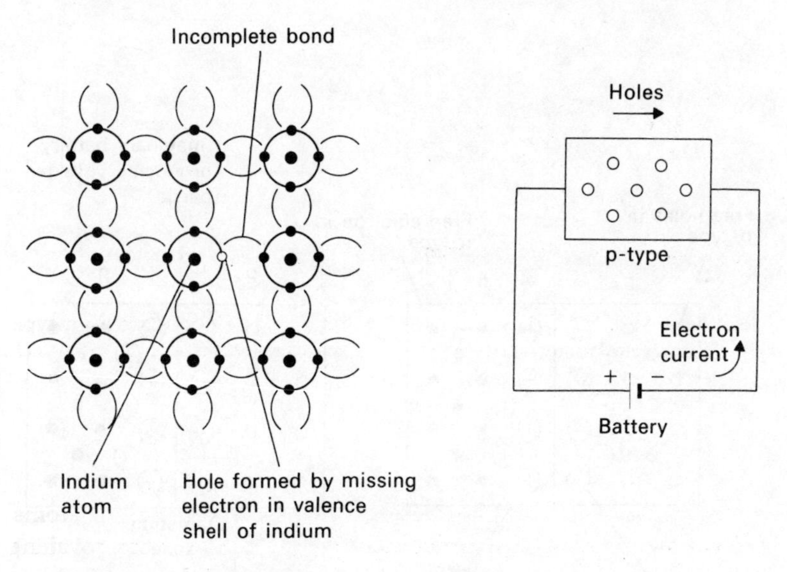

Fig. 13/6. P-type semiconductor.

Fig. 13/7. Conduction in p-type semiconductors.

Conclusions. In n-type semiconductors current flow is due to electrons. In p-type semiconductors current flow is due to holes.

P-N JUNCTION

When a p-type and an n-type semiconductor are joined together an interesting effect is noticed (Fig. 13/8).

Some of the free electrons from the n-type wander across the junction into the p-type, and similarly some of the holes from the p-type wander across the junction into the n-type. The electrons that have wandered into the p-type combine with some of the holes and form negative ions. Similarly some of the holes which have wandered across into the n-type combine with some of the free electrons and form positive ions, i.e. the previously neutral p- and n-type materials become charged.

This movement of electrons and holes continues until a charge builds up on either side of the junction, which then repels any further electrons or holes from crossing the junction.

We can consider this charge barrier as being an imaginary battery (Fig. 13/9).

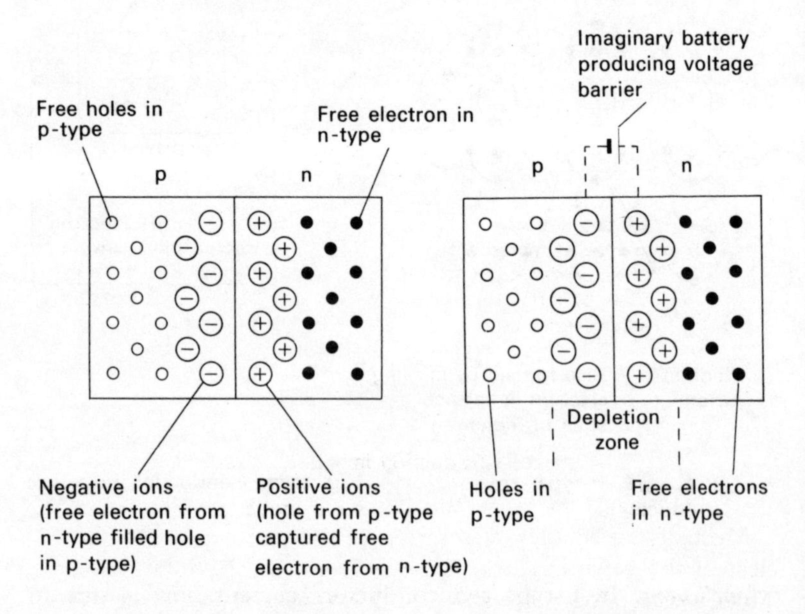

Fig. 13/8. P–N junction. **Fig. 13/9.**

The region in the vicinity of the junction where there are now no free electrons or holes is known as the depletion zone or depletion layer. It acts as an almost perfect insulator.

To permit a continuous flow of current we must overcome this imaginary battery or voltage barrier, since it is preventing further movement of electrons and holes.

This can be achieved by connecting an external battery, with its negative terminal to the n-type semiconductor and its positive terminal to the p-type, i.e. it is connected to *overcome* the voltage barrier set up across the junction, leaving only a very small depletion zone around the junction.

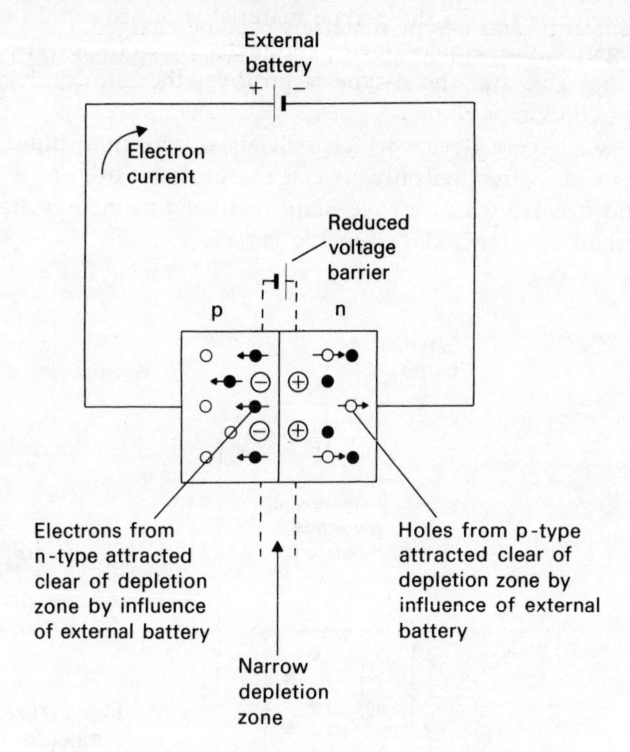

Fig. 13/10. Conduction in p-n junction.

Most of the charges now crossing the junction are attracted clear of the depletion zone by the effect of the external battery. Thus, a continuous stream of electrons is able to flow through the semiconductors. Electrons and holes which move through the semiconductor are replaced by electrons and holes from the battery.

Now let us consider what happens when the external battery is connected the other way round, i.e. with the positive terminal to the n-type semiconductor and the negative terminal to the p-type.

This time the external battery is connected in such a way as to reinforce the voltage barrier, making the depletion zone wider and therefore preventing any flow of electrons or holes across the barrier. Therefore, no current can flow through the semiconductors.

Conclusion. When the p-type material is positive and the n-type negative, the semiconductor conducts. When the p-type material is negative and the n-type is positive, the semiconductor does not conduct.

We have a device which will allow current to flow through it in one direction only. It can therefore be used as a rectifier, and it is known as a solid-state rectifier or junction diode. The circuit symbol is shown in Fig. 13/12.

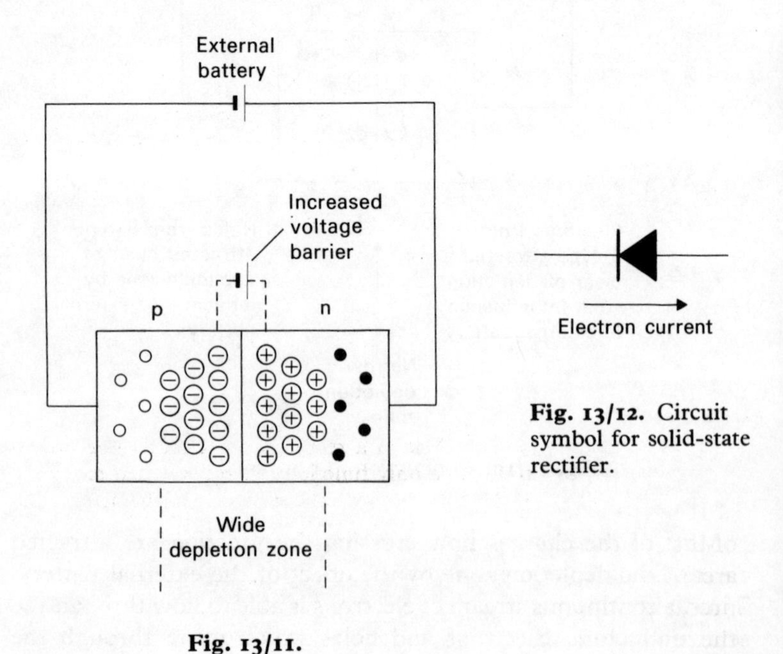

Fig. 13/11.

Fig. 13/12. Circuit symbol for solid-state rectifier.

ADVANTAGES OF SOLID-STATE RECTIFIERS

Solid-state rectifiers:
(a) do not require any filament heating circuit
(b) are robust
(c) have long life
(d) are cheap
(e) do not face any danger of producing x-rays (in some circumstances diode valves could emit x-rays).

RECTIFICATION IN X-RAY SETS

SELF-RECTIFIED CIRCUIT

We mentioned in Chapter 12 that an x-ray tube is a modified diode valve. Therefore it acts as a rectifier in its own right. Thus, in the circuit in Fig. 13/13 it will only allow current to flow in the direction of the arrow, namely cathode to anode.

The voltage and current wave forms are shown in Fig. 13/14.

The current is rectified because the anode end of the x-ray tube is not hot and does not, therefore, emit electrons—*but* is this true?

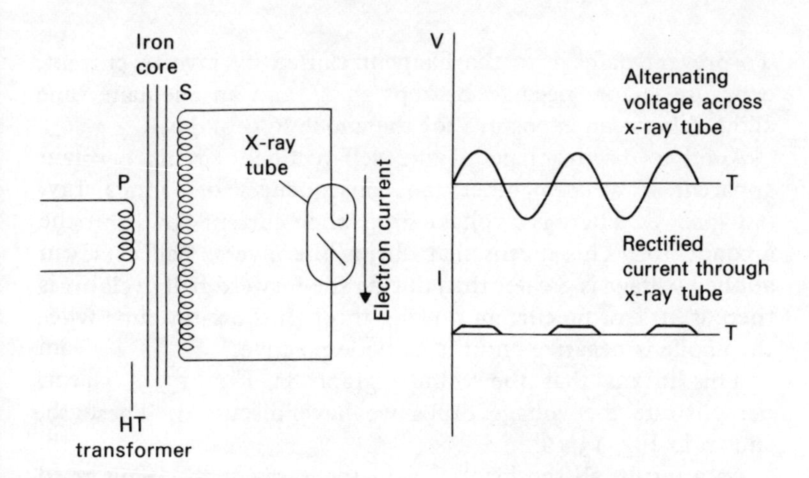

Fig. 13/13. Self-rectified circuit.

Fig. 13/14. Voltage and current in self-rectified circuit.

We said in Chapter 12 that 99% of the energy put into an x-ray tube appears as heat. Therefore, the anode end of the x-ray tube will get very hot and will, therefore, emit electrons during the half cycle when it is negative. Since the cathode will at that moment be positive these electrons will bombard the filament. Because the filament is only a thin piece of wire it would soon be destroyed by this attack.

Fig. 13/15 is a graph of current flowing through the x-ray tube showing the presence of inverse current.

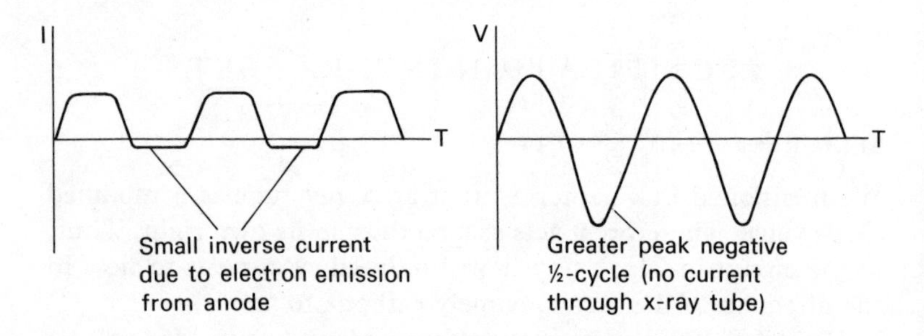

Fig. 13/15. Current in self-rectified circuit showing inverse current.

Fig. 13/16. Voltage across x-ray tube in self-rectified circuit showing greater inverse voltage.

To prevent damage to the filament caused by inverse current, exposure factors need to be kept small, and an adequate time allowed between exposures for the anode to cool down.

Another disadvantage of the self-rectified circuit becomes apparent if we remember the consequences of Ohm's Law (Chapter 7). There is a voltage drop when current flows through a conductor. This means that during the inverse half cycle the applied voltage is greater than during the forward half cycle since there is little or no current flowing through the x-ray tube when the anode is negative and the cathode positive.

This means that the voltage graph in Fig. 13/14 should demonstrate the voltage drops we have discussed. These are shown in Fig. 13/16.

As a result, all the insulation in the x-ray tube circuit must be made to withstand greater voltages than are used in the production of x-rays.

FULL-WAVE-RECTIFIED CIRCUIT

By using four solid-state rectifiers connected in the form of a bridge circuit it is possible to use both the forward and inverse parts of the voltage cycle. Fig. 13/17 shows such a bridge circuit connected between the high tension transformer and the x-ray tube.

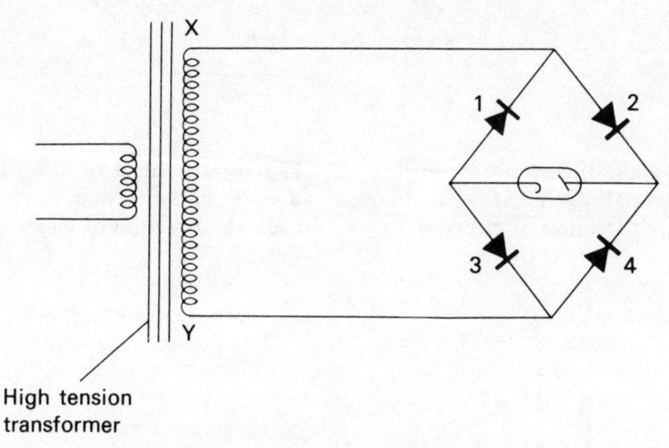

Fig. 13/17. Full-wave rectified circuit.

X and Y are the output leads from the high tension transformer providing alternating voltage.

1. Let us look first at what happens when X is negative and Y is positive (Fig. 13/18). Electrons move from X to 1 to x-ray tube to 4 to Y (attracted by positively-charged Y).

You will notice that the electrons can only pass through the x-ray tube; other routes are blocked by rectifiers 2 and 3.

2. Now let us look at the next half cycle, when X is positive and Y is negative (Fig. 13/19). Electrons move from Y to 3 to x-ray tube to 2 to X (attracted by positively charged X), i.e. electrons pass through the x-ray tube in the same direction during the second half cycle.

If we now look at graphs of voltage and current in different parts of the circuit we see that current flows through the x-ray tube during every half cycle.

If we now incorporate the voltage and current across the x-ray tube into one graph we can compare it with that for the self-rectified circuit.

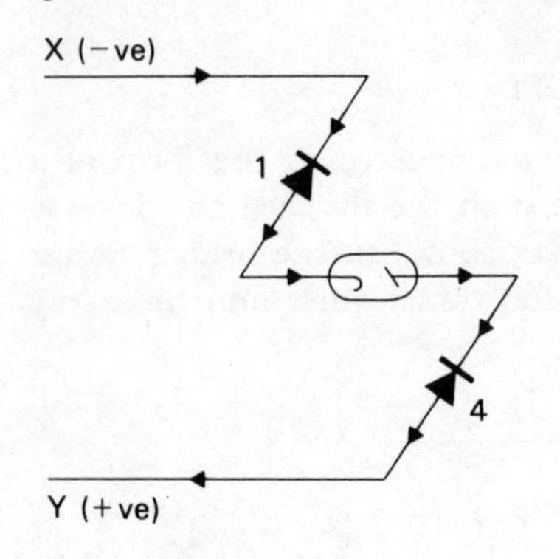

Fig. 13/18. Electron flow during the first half cycle. Arrows indicate direction of electron flow.

X (+ve)
2
3
Y (−ve)

Fig. 13/19. Electron flow during the second half cycle. Arrows indicate direction of electron flow.

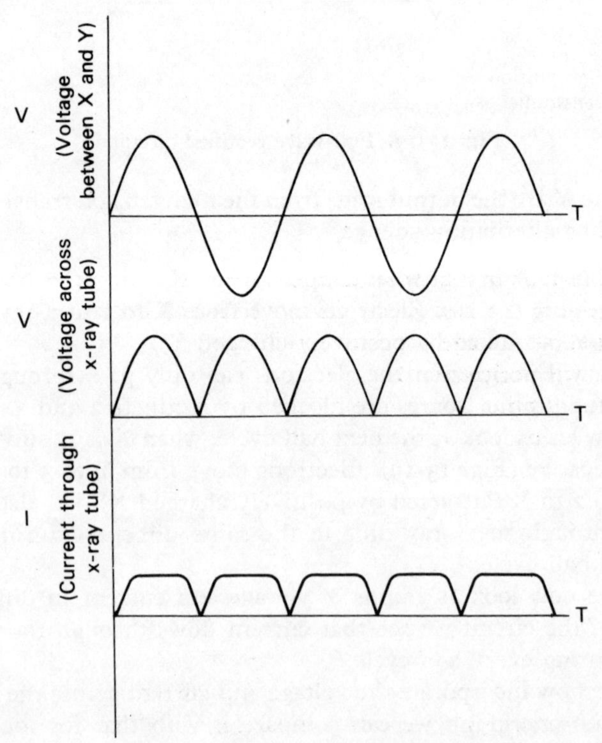

Fig. 13/20. Voltage and current graphs at points in the full-wave circuit.

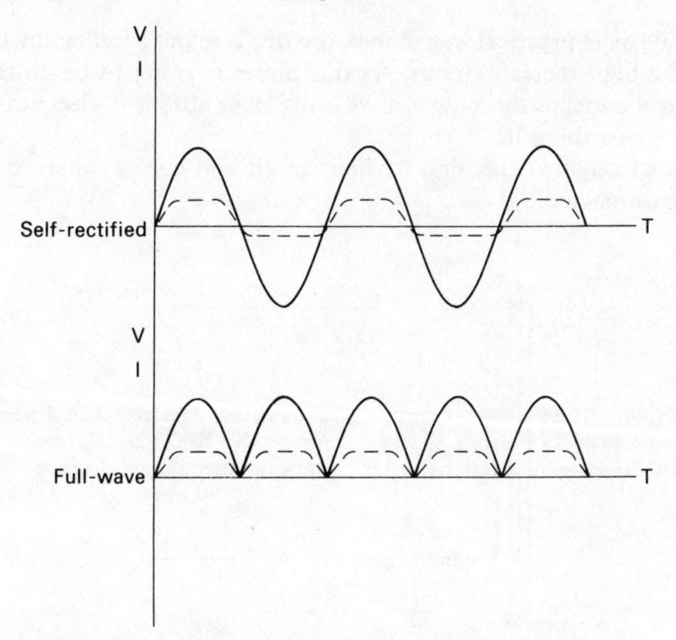

Fig. 13/21. Comparison of voltage and current graphs in self- and full-wave rectification. Solid line, voltage; dotted line, current.

ADVANTAGES OF FULL-WAVE OVER SELF-RECTIFICATION

1. More x-rays are generated in a given time because current flows through the x-ray tube during both half cycles.
2. No inverse voltage occurs, therefore,
  (a) no inverse current
  (b) no filament bombardment
  (c) no need to allow anode to cool
  (d) higher mA values can be used.

This is the circuit that is used in most medium power x-ray equipment.

THE mA METER

As you will see in Chapter 16, the amount of x-rays produced is proportional to the tube current (mA).

The most practical way to measure this is to put a milliammeter in the high tension circuit. As this meter is going to be on the control panel of the x-ray set we must make sure it is electrically safe by earthing it.

Let us look at the high tension circuit and decide where it is best connected.

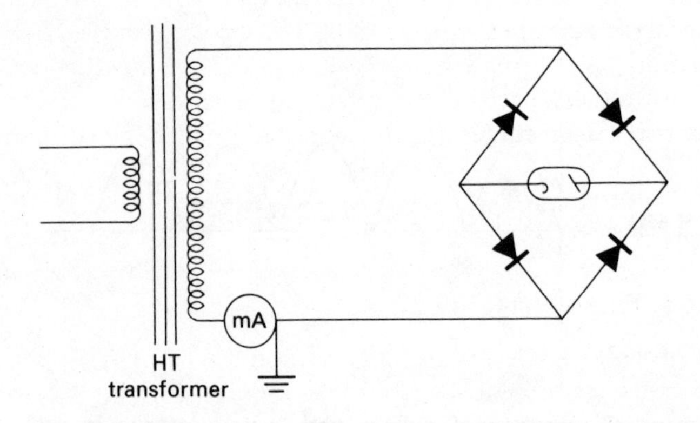

Fig. 13/22. Choosing a position for the mA meter (choice 1).

If point Y is earthed it will always have zero potential. Thus, for a given exposure of 100 kVp point X will vary between 100 kV (+ve) and 100 kV (−ve) through the cycles of a.c.

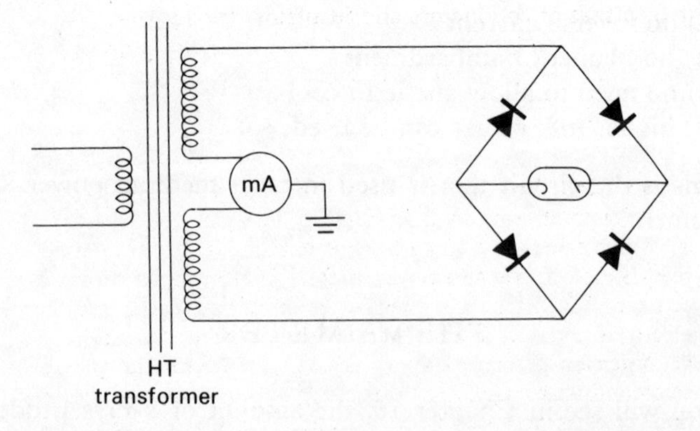

Fig. 13/23. Choosing a position for the mA meter (choice 2).

If we earth the centre point of the secondary winding of the HT transformer we effectively split the secondary into two parts. This time the centre point is at zero potential and for the same exposure as in choice 1, namely 100 kVp, we find that as the potential at point X varies from 50 kV (+ve) to 50 kV (−ve); simultaneously, the potential at point Y varies from 50 kV (−ve) to 50 kV (+ve); i.e. the *difference* in potential (across the x-ray tube) is still 100 kVp but the maximum voltage at any point in the secondary circuit is only 50 kVp. This means that the electrical insulation can be reduced since it has only to insulate against half the voltage at which the x-ray tube operates.

For this reason choice 2 is the one used in all x-ray equipment.

SUMMARY

The reasons for earthing the centre of the secondary winding of the HT transformer are as follows:

1. It provides a safe point of connection for the mA meter.
2. It splits the high voltage, thereby, reducing the maximum voltage at any point in the circuit by half.

CHAPTER SUMMARY

1. Semiconductors have a resistance whose value lies between that of conductors and insulators (p. 140).
2. Their resistance can be altered by incorporating additives (p. 140). N-type semiconductors contain additives (e.g. antimony) which have five electrons in their valence shell, leaving electrons free to conduct current (p. 141).
3. P-type semiconductors contain additives (e.g. indium) which have three electrons in their valence shell leaving 'holes' available to conduct current (p. 142).
4. When a p-type and an n-type semiconductor are 'fused' together electrons and holes in the vicinity of the junction combine, forming a depletion zone where there are no electrons or holes available to conduct current (p. 144).
5. When an external battery is connected with the p-type semiconductor positive and the n-type negative, the depletion zone is reduced and current will flow through the material (p. 145).

6. When an external battery is connected with the p-type semiconductor negative and the n-type positive, the depletion zone is increased and current cannot flow through the material (p. 146).
7. In the self-rectified circuit the x-ray tube acts as a rectifier and only a half the a.c. cycle is used to produce x-rays (p. 147).
8. In a full-wave circuit four rectifiers ensure that the voltage is always in the forward direction through the x-ray tube and both halves of the a.c. cycle are used to produce x-rays (p. 149).
9. The mA meter is used to measure the current through the x-ray tube during exposures. It is connected to the centre point of the secondary winding of the high tension transformer and earthed for safety (p. 152).

Chapter 14
Electromagnetic radiations

Having studied the constructional aspects of the x-ray tube and its electrical supplies we shall now concentrate on the radiations it produces. X-rays are one example of a whole range of radiations known as electromagnetic radiations.

Whenever electric charges accelerate or decelerate disturbances are set up in the electric and magnetic fields around them. We say an electromagnetic wave has been generated. A useful, and widely quoted analogy is to imagine an object floating on perfectly calm water. When the object is moved, disturbances are produced in the water which cause surface waves to be generated which carry the disturbances outwards and away from the source (Fig. 14/1). Other objects floating nearby would experience these surface waves and be disturbed by them. We may infer that to produce this effect, the surface waves must be carrying energy. Electromagnetic waves also carry energy.

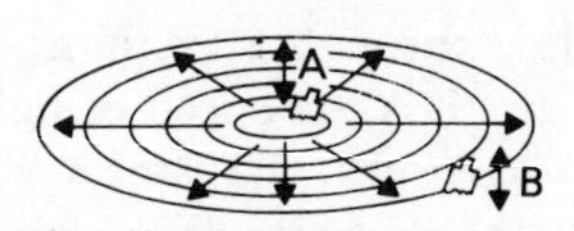

Fig. 14/1. Two bottles floating on still water. Disturbances to bottle A are passed on to bottle B by surface waves on the water; i.e. energy is transmitted from A to B.

ENERGY AND FREQUENCY OF ELECTROMAGNETIC WAVES

The quantity of energy carried by electromagnetic waves depends on the frequency of repetition of the disturbances in the electric and magnetic fields. We define the frequency as being the number of waves experienced per second. It is usually measured in 'cycles' per second (c.p.s.) and the SI unit of frequency is the hertz (Hz), where 1 Hz = 1 cycle per second. (We have come across this term before in Chapter 10 in relation to alternating current.)

Wave energy (E) is proportional to wave frequency (f); i.e. $E \propto f$. It therefore follows that high frequency waves carry more energy than low frequency waves.

ELECTROMAGNETIC SPECTRUM

Electromagnetic waves can be produced by various means resulting in different values of frequency. Because they have different frequencies they also have different energy values and different methods are required in order to detect them. The complete range of frequencies and energies is called the electromagnetic spectrum and is illustrated in Table 14/1. Radio waves are at one end of the spectrum, having comparatively low frequencies and carrying little energy, while at the other end cosmic rays have very high frequencies and are very energetic. Visible light rays have medium-value frequencies, while x- and gamma rays lie towards the high energy end of the spectrum. All these electromagnetic radiations travel very quickly; at roughly 186 000 miles per second or the 'speed of light'. At this speed radio waves can travel to the moon and back in less than three seconds. The symbol c is used universally to represent the speed of electromagnetic radiation. In SI units $c = 3 \times 10^8$ m/s, which is the speed of electromagnetic waves when travelling through a vacuum; in other media the speed is reduced, e.g. in glass the speed of light is 2×10^8 m/s. All electromagnetic waves travel at the same speed in a vacuum, and conveniently, their speed through air is almost exactly the same.

WAVELENGTH

Considering our analogy with the surface waves on water, we can see that successive disturbances are separated by a constant distance which is known as the wavelength. The same is also true of the electromagnetic waves: the distance between successive electromagnetic disturbances is the wavelength of the radiation. The Greek letter lambda (λ) is employed to denote wavelength.

Table 14/1. The spectrum of electromagnetic radiations ranging from high energy (short wavelength) cosmic rays, to low energy (long wavelength) radio waves.

Radiation		Source
Cosmic rays		Outer space
Gamma rays		Radioisotopes
X-rays		X-ray tubes
Ultra-violet (UV) rays		The sun, UV lamps
Visible light	violet indigo blue green yellow orange red	the sun, incandescent and fluorescent lamps
Infra-red (heat) rays		the sun, infra-red lamps hot objects
Microwaves		microwave ovens
Radar waves		radar transmitters
TV and radio waves	UHF (ultra high frequency) VHF (very high frequency) Short wave Medium wave Long wave	TV and radio transmitters

Let us suppose that one disturbance is arriving at a particular point every second; i.e. the frequency, $f = 1$ Hz, travelling at a speed of c metres per second. Then it follows that the disturbances will be separated by a distance of c metres; i.e. the wavelength is c metres.

If two disturbances are arriving each second ($f = 2$ Hz) also at a speed of c metres per second, then the wavelength must be $c/2$ metres.

Finally, if f disturbances arrive every second (i.e. frequency is f), then the wavelength (λ) must be c/f;

$$\text{i.e.} \quad \lambda = \frac{c}{f}$$

$$\text{or} \quad c = f\lambda.$$

This relationship between the speed, frequency and wavelength of a wave holds true for all wave systems, not just for electromagnetic radiations.

Because the speed of electromagnetic waves is a constant (c), radiations with a high frequency will have a short wavelength and those with a low frequency will have a long wavelength. Knowing the value of frequency we can calculate the wavelength and vice versa. For this reason it is not normally necessary to quote both the frequency and wavelength of radiation. Sometimes it is more convenient to specify the wavelength of electromagnetic radiation than its frequency. Until recently this was the traditional method for long- and medium-wavelength radio transmissions, but the recent trend is towards quoting the frequency; e.g. BBC Radio 4 UK has a wavelength of 1500 metres and a frequency of 200 kilohertz (200 kHz). Remembering $c = f\lambda$ we can see that in this case

$$c = 200 \times 10^3 \times 1500$$
$$= 3 \times 10^8 \text{ m/s}$$

which reinforces our previous knowledge of the speed of electromagnetic radiations.

The wavelength of long wave radio transmissions is almost 1 mile, while VHF (very high frequency) radio waves have wavelengths of only a few centimetres. The wavelength of visible light is much shorter, being measured in hundreds of nanometres (1 nm = 10^{-9} m). The wavelengths of x- and gamma rays are measured in minute fractions of nanometres (e.g. 100-kV x-rays have a wavelength of about 0.01 nm). Prior to the introduction of SI units, a unit of length called the ångström (Å) was widely used for measurements of small radiation wavelengths:

$$1 \text{ ångström unit} = 10^{-10} \text{ m}$$
$$= 0.1 \text{ nm.}$$

RECTILINEAR PROPAGATION OF RADIATION

Electromagnetic rays travel in straight lines; i.e. they exhibit rectilinear propagation. They may deviate when passing through a junction between one medium and another; e.g. light is bent (refracted) on passing from air to glass, as in an optical lens. Refraction is *not* apparent with x- and gamma rays and for all practical purposes we may assume that these radiations travel in straight lines without deviation.

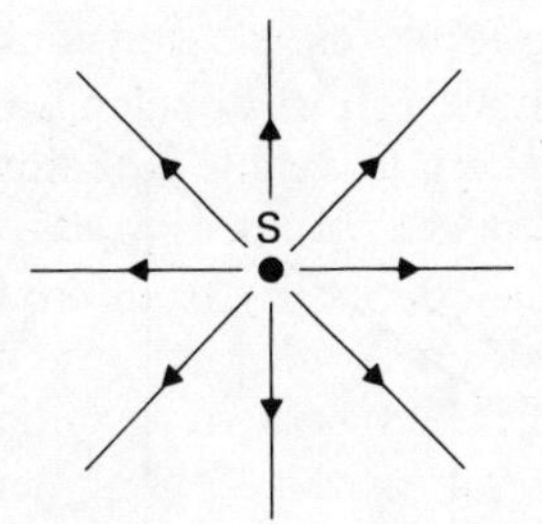

Fig. 14/2. Divergent electromagnetic rays being emitted from a source S.

Because electromagnetic rays travel in straight lines it follows that if they are generated from a point source (S in Fig. 14/2) they will diverge, causing a reduction in the intensity of the radiation (i.e. in the concentration of radiation energy) as the distance from the source is increased. In other words, as we move further from the source of radiation, its effects become weaker. The exact way in which this reduction in intensity occurs is defined in the Inverse Square Law which we first described in Chapter 3 in relation to electric field strength. The Inverse Square Law follows directly from the fact that radiation travels in straight lines.

INVERSE SQUARE LAW FOR RADIATION

Fig. 14/3 illustrates a point source (S) from which electromagnetic radiation is being emitted at a constant rate. If we consider a particular 'window' of area A_1 at a distance d_1, through which the radiation is passing, we can imagine that a certain quantity of radiation energy, E, passes through every second. That same energy must also pass through the larger window of area A_2 at a distance d_2 from the source. No energy can be 'lost' or 'gained' between the two windows because the radiation is constrained to travel in straight lines originating from S. The concentration of energy passing through A_2 is less than through A_1 because the same energy is passing through a larger window area.

Radiation intensity (I) is defined as the quantity of radiation energy passing through a unit area in unit time;

i.e. it is $I_1 = \dfrac{E}{A_1}$ at the near window. But area $A_1 \propto d_1^2$

$$\text{so} \quad I_1 \propto \frac{1}{d_1^2}.$$

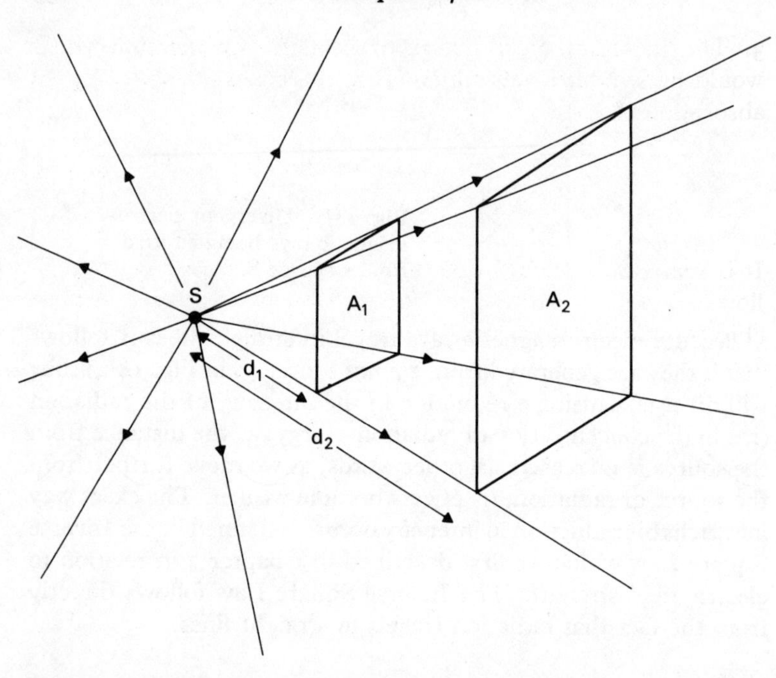

Fig. 14/3. The same radiation energy passes through A_2 as passes through A_1, but its concentration is less at A_2 because it is spread over a larger area. The area of each window is proportional to the square of its distance from the source.

Similarly, $I_2 = \dfrac{E}{A_2}$ at the far window. But area $A_2 \propto d_2^2$

$$\text{so} \quad I_2 \propto \frac{1}{d_2^2}$$

i.e. radiation intensity is inversely proportional to the square of the distance from the source of radiation.

This is the Inverse Square Law as it applies to electromagnetic radiations. We must remind ourselves, however, that it only applies under certain conditions:

1. The source must be a point. (In practice it must be very small compared with the distance at which intensity is being considered.)

2. The emission of energy from the source must be distributed equally in all directions.

3. The radiation must not pass through any medium which would cause additional reductions in the beam intensity, e.g. by absorption of energy.

QUANTUM THEORY

It is sometimes useful to think of radiation not as a continuous flow of wavelike disturbances but as a stream of separate particles (Fig. 14/4). The Corpuscular Theory of radiation was based on

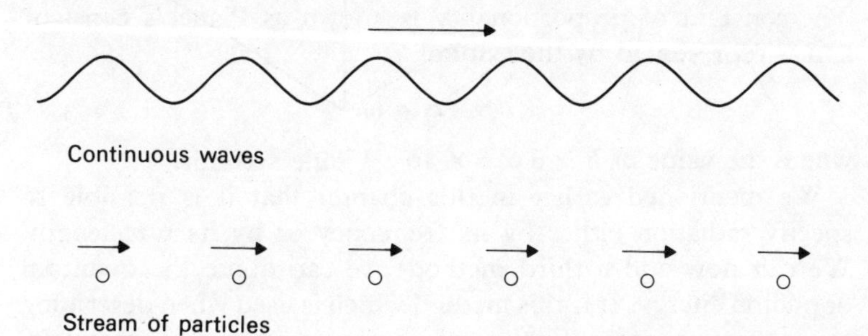

Fig. 14/4. The wave and corpuscular ideas of radiation.

this idea. However, there are also times when the wave properties of electromagnetic radiation are more relevant and the solution is therefore to try to combine the advantages of both the corpuscular and wave concepts into a single all-embracing theory, the Quantum Theory, which pictures the wave-like disturbances being produced in a succession of individual bursts of energy called quanta or photons (Fig. 14/5).

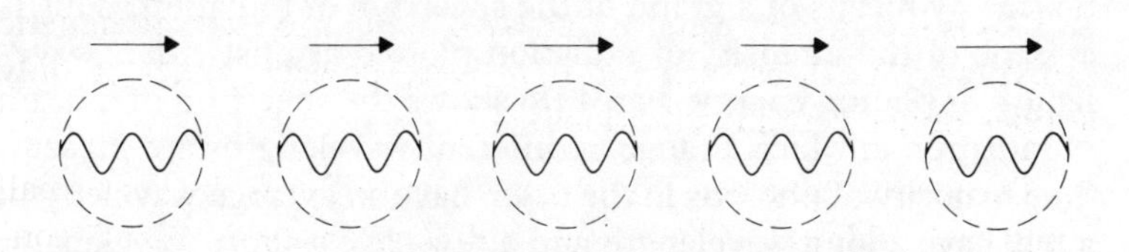

Fig. 14/5. The Quantum Theory of radiation.

These quanta have no mass or electric charge, but consist entirely of pure energy. They have the properties of waves (e.g. wavelength and frequency) as well as properties of particles (e.g. they may rebound or be scattered if they collide with particles such as electrons). Quanta also have the property of being indivisible; they cannot be split into smaller units.

The energy (E) of a quantum is proportional to its wave frequency (f):

$$E \propto f$$

$$\text{or} \quad E = \text{constant} \times f$$

The constant of proportionality is known as Planck's constant and is represented by the symbol $\hbar$,

$$\text{i.e.} \quad E = \hbar f$$

where the value of $\hbar = 6.626 \times 10^{-34}$ joule-seconds.

We mentioned earlier in this chapter that it is possible to specify radiation either by its frequency or by its wavelength. We can now add a third method: we can quote the quantum or photon energy. It is this method which is used when describing x- or gamma rays. The energy units used are electronvolts (see Chapter 5) rather than joules; e.g. the photon energy of x-rays used for chest radiography might be 55 keV, and for orthovoltage radiotherapy might be 250 keV.

SPECTRAL CURVES

A particular beam of electromagnetic radiation may well contain a mixture of photons having different values of photon energy. It is useful in such cases to be able to show the range of the different photon energies and their relative intensities. We can do this by means of a graph of the spectrum of radiation. This is a graph of the intensity of radiation plotted against either wavelength or photon energy. Fig. 14/6 shows the spectrum of a beam of radiation in which a range of different wavelengths are present. The majority of photons in the beam have an average wavelength, a few have a long wavelength and a few have a short wavelength. Fig. 14/7 shows the corresponding graph of intensity plotted against photon energy. This tells us that in this particular beam most of the photons have an average value of energy, a few have

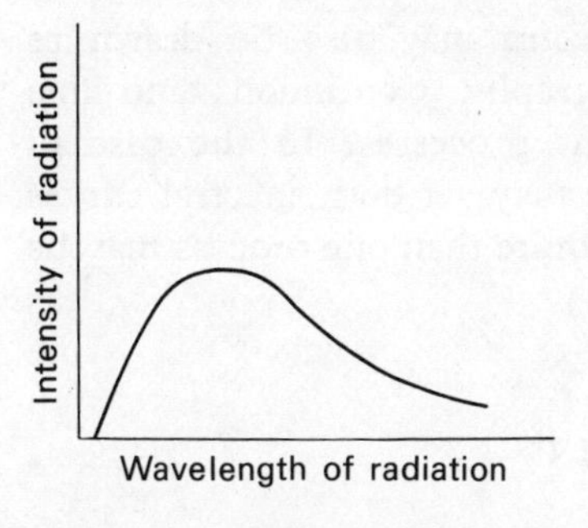

Fig. 14/6. An example of a continuous spectrum of radiation drawn as a wavelength spectrum.

Fig. 14/7. The photon energy spectrum corresponding to the spectrum shown in Fig. 14/6.

high energy and a few low energy. Remembering that a high energy photon has a short wavelength and a low energy photon has a long wavelength we can see that the two forms of the graph are similar, but not identical.

The spectral curves shown in Figs. 14/6 and 14/7 describe a beam of radiation in which many different photon energies are present with no gaps or sudden peaks. The graph is a smooth curve and is known as a continuous spectrum. In some circumstances it is possible to produce a beam containing only a few particular photon energies. The spectral curve for this type of beam is shown in Fig. 14/8 and is known as a line spectrum or

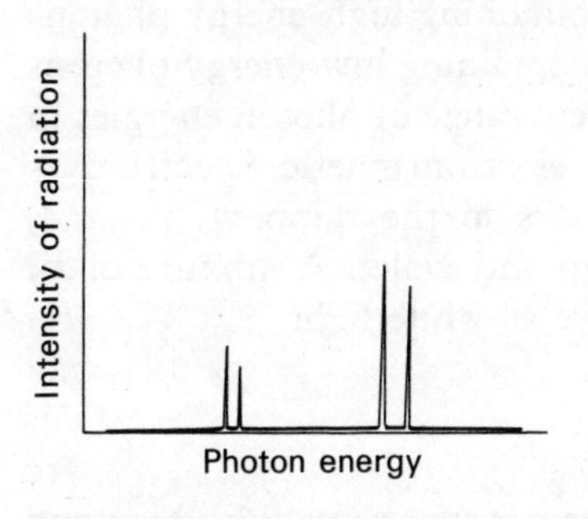

Fig. 14/8. An example of a line or characteristic spectrum of electromagnetic radiation showing photons present at four energy levels.

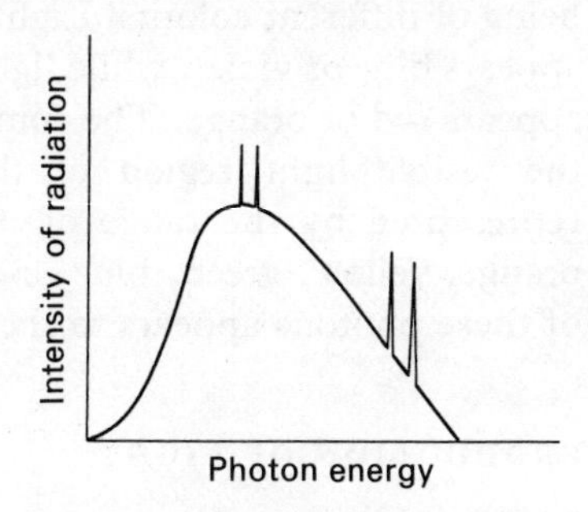

Fig. 14/9. An example of a line spectrum superimposed on a continuous spectrum.

characteristic spectrum. Line spectra may also be drawn as wavelength rather than energy graphs. Continuous and line spectra are produced by different processes. In the case of x-radiation from an x-ray tube we may get both spectral curves superimposed (Fig. 14/9) because more than one process may be involved in producing the x-rays.

LIGHT

Before describing the properties of x- and gamma radiation we shall spend a short time investigating some of the properties of visible light. We will then find it easier to understand the processes which are involved in the production of x-rays.

INTENSITY OF LIGHT

In relation to visible light, intensity may be thought of as brightness; e.g. a bright light is of high intensity. Intensity is still strictly the energy falling on a surface of unit area in unit time.

COLOUR OF LIGHT

Photons of light which have different energies, wavelengths and frequencies affect the human eye differently and we see them as being of different colours. Light containing high-energy photons appears blue or violet, while light containing low-energy photons appears red or orange. The complete range of photon energies in the visible light region of the electromagnetic spectrum is represented by the range of colours in the rainbow; i.e. red, orange, yellow, green, blue, indigo and violet. A mixture of all of these photons appears to the eye as white light.

PRODUCTION OF LIGHT

Light may be generated in two main ways, namely by heat and by electron transition.

Heat. If a solid is heated sufficiently, its atoms and molecules vibrate causing photons of electromagnetic radiation (infra-red rays) to be emitted. If the temperature reaches a high-enough level some of the photons will be of visible light. We say the object is 'red hot', meaning photons of red light are being emitted. As the temperature is increased, photon energies will rise and photons of yellow, green, blue and even violet light will be produced. If all of these are present the eye would perceive the mixture as white light and we would say the object was 'white hot'. Fig. 14/10 shows the spectrum of light emitted by an object at 400°C, 700°C and 1500°C. We can recognise these as forms of the continuous spectrum showing that at 400°C no

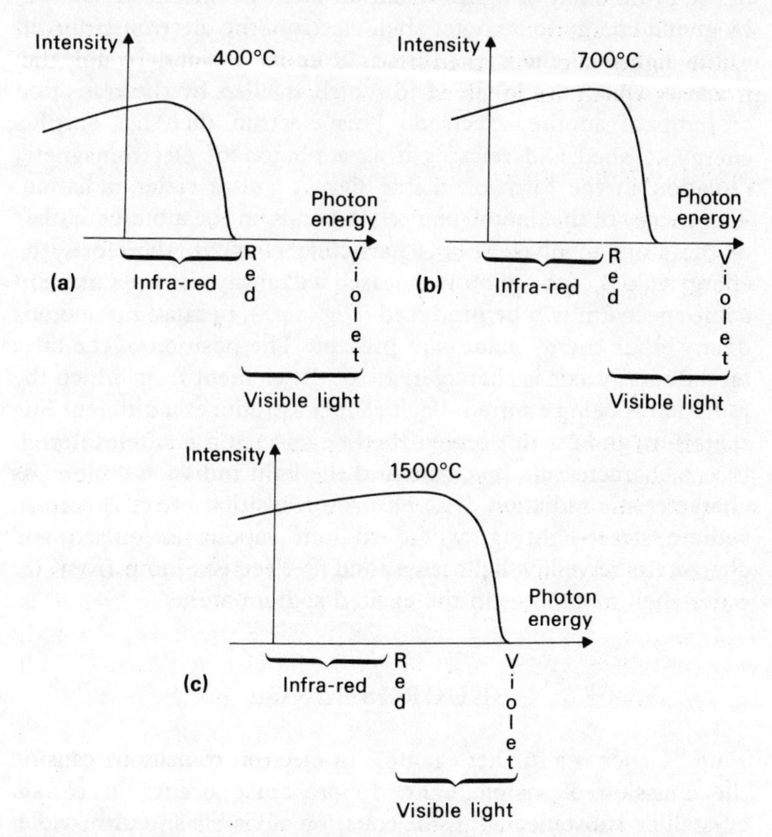

Fig. 14/10. Radiation spectra produced at three different temperatures. (a) At 400°C the body emits infra-red rays but no visible light; (b) at 700°C the body emits visible light from the red end of the spectrum as well as infra-red rays; (c) at 1500°C the body emits light of all energies as well as infra-red rays—it glows 'white hot'.

visible light is produced, only invisible heat rays (infra-red). At 1500°C all the colours of the spectrum are emitted and white light results. At the intermediate temperature, 700°C, some visible light is emitted as well as infra-red rays, and we see the object glowing red.

It is this effect, known as *incandescence,* which causes conventional electric light bulbs, gas lamps and candle flames to emit light.

Electron transition. This is a different process which also results in the production of light. When an atom is excited or ionised, by giving energy to an outer shell electron, the electron is forced to leave its shell and move further from its nucleus. In doing so it leaves a vacancy in the shell which is filled by the transition or jump of another electron. This electron then has surplus energy to shed and releases it as a photon of electromagnetic radiation in the form of visible light or ultra-violet radiation. The energy of the photon emitted depends on the atomic number of the atom involved. For a particular element, therefore, the energy values of the photons released will always be the same and a line spectrum will be produced (Fig. 14/8), because no photons of any other energy values are present. The position of the lines on the energy axis is characteristic of the element from which the radiation is being emitted. Each element produces a different line spectrum, and for this reason the line spectrum is often referred to as a characteristic spectrum and the light radiation emitted as characteristic radiation. The electron transition effect is seen in sodium street-lighting, where sodium vapour is ionised and characteristic yellow light is emitted as electrons jump from one outer shell to another in the excited sodium atoms.

FLUORESCENCE

Fluorescence is a further example of electron transitions causing the emission of visible light. Fluorescence occurs in certain crystalline substances. It is the emission of visible and ultra-violet light from a substance following the absorption of higher energy (shorter wavelength) radiation such as x-rays, or energetic particles such as high speed electrons. Again the photon energies emitted are characteristic of the elements forming the crystals.

The exposure to high energy radiation causes atoms in the crystal to become excited. Some of the electrons in outer shells move out of their orbits to a region further from the nucleus. As they fall back again they release photons of radiation in the form of light or ultra-violet rays. This effect is useful in radiography because it offers a method of converting x-ray energy, which is invisible to the eye and to which photographic film is relatively insensitive, into visible light which the eye can detect and to which films are very sensitive.

Examples of fluorescence are:

1. X-ray intensifying screens which employ fluorescent materials such as calcium tungstate to convert x-rays into a blue-violet light which is then used to produce an image on 'screen' film.
2. Television screens carry a coating of fluorescent material which emits light when excited by a beam of high speed electrons inside the TV tube.

In some fluorescent materials, the return of the electrons to their original shells may continue for some time after the exposure to x-rays has ended. This effect is known as *phosphorescence* or *afterglow*. A limited degree of afterglow is an advantage in intensifying screens, but in TV screens which carry moving images it is not acceptable.

From the work we have studied so far in this unit, we know that light has two important effects:

1. The human eye responds to it, i.e. we can see light.
2. Photographic film responds to it, i.e. light causes chemical changes to occur in photographic emulsion which can be used to record images.

There are other effects such as photosynthesis in green plants, and the synthesis of vitamin D in human skin. While these are not appropriate for studying here, an effect known as photoelectric emission is, however, relevant.

PHOTOELECTRIC EMISSION

Some materials (e.g. selenium, caesium and lithium) possess electrons which may be released from their atoms merely by exposing the material to light. Under suitable conditions it may be possible to measure the release of electrons and use this to determine the intensity of light to which the material was exposed. The exposure meter used with photographic cameras is an example of the use of this effect.

A device used to convert light into electrical energy is generally called a *photocell*. In one form of photocell the effect of the light is to alter the electrical resistance of the device, which is known as a *photoresistor* and employs semiconducting materials which respond to variations in intensity of light. A photoresistive device is used with an external supply of electrical power.

In another form the semiconductor in the photocell generates a potential difference when exposed to light and does not require an external electrical supply to operate.

In the x-ray set photocells may be used to terminate an x-ray exposure after the correct amount of light from a fluorescent screen has exposed the film. This form of exposure control is known as a *phototimer*.

We have now completed our description of the general behaviour of electromagnetic radiations, and of some of the properties of visible light. In the next chapter we begin our study of the high energy radiation produced by x-ray tubes on which much of the work of a radiographer relies.

CHAPTER SUMMARY

1. Electromagnetic radiation is generated when electric charges accelerate or decelerate (p. 155).
2. Electromagnetic waves carry energy (p. 155).
3. Frequency of radiation is the number of waves experienced per second. It is measured in hertz (p. 156).
4. The complete range of electromagnetic radiations from low energy to high energy is known as the spectrum (p. 156).
5. All electromagnetic waves travel at the same constant speed through a vacuum (p. 156).
6. Wavelength is the distance between successive electromagnetic disturbances (p. 157).
7. Electromagnetic waves travel in straight lines (p. 159).
8. Quanta or photons are the minute individual pulses of energy of which radiation consists (p. 161).
9. Radiation spectra may be continuous, characteristic or superimposed (p. 163).

10. Light may be generated by heat or by electron transition (p. 164).
11. Characteristic radiation is produced when electrons change shells (p. 166).
12. Fluorescence is an example of the production of light by electron transition (p. 166).
13. Photoelectric emission is the release of electrons from a material on exposure to light (p. 167).

KEY RELATIONSHIPS

$c = \lambda f$ (wavelength) (p. 158).

$I \propto \frac{1}{d^2}$ (radiation intensity) (p. 159).

$E = hf$ (Planck's constant) (p. 162).

Chapter 15
X-rays

We are now in a position to be able to study x-rays: how they are produced and the properties they possess.

PRODUCTION OF X-RAYS

X-rays are produced when electrons lose a lot of energy. The electron energy may be lost by the slowing down (deceleration) of fast moving electrons or by electron transitions between the inner shells of an atom. In the x-ray tube both of these processes are involved.

The x-ray tube is a device which is designed to produce fast-moving electrons and to cause them to decelerate rapidly. In Chapter 12 we studied the construction of the x-ray tube in detail. Let us now remind ourselves of the basic features illustrated in Fig. 15/1. The tube contains:

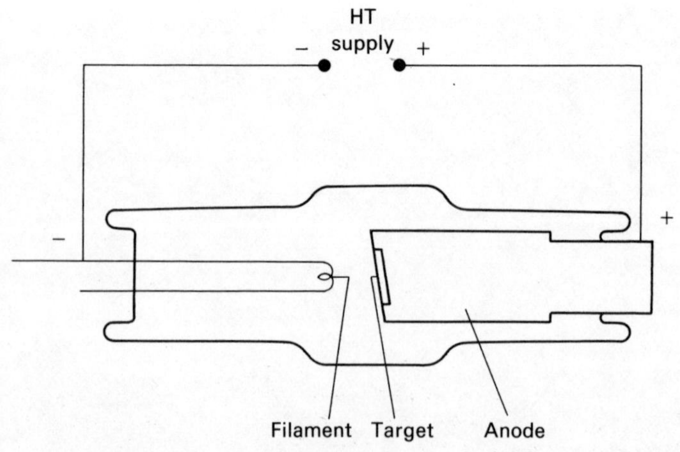

Fig. 15/1. Some basic features of the x-ray tube.

1. A heated filament which releases negative electrons.
2. A positive anode which attracts them.
3. A high tension supply which accelerates the electrons to very high speeds.
4. A tungsten target (part of the anode) whose job is to slow down the fast moving electrons very rapidly, thus causing x-rays to be emitted.

What actually happens at the target? The electrons interact with the atoms of the target material in several different ways, some of which cause x-rays to be produced.

INTERACTIONS AT THE TARGET

Four types of interaction can take place:

1. The incoming electron from the filament enters the target and transfers a fraction of its energy to an outer shell electron of an atom in the target, pushing it out to a higher energy level. The electron then falls back to its original level and releases its surplus energy as heat in the target. (The original electron may go on to repeat this process many times; see Fig. 15/2).

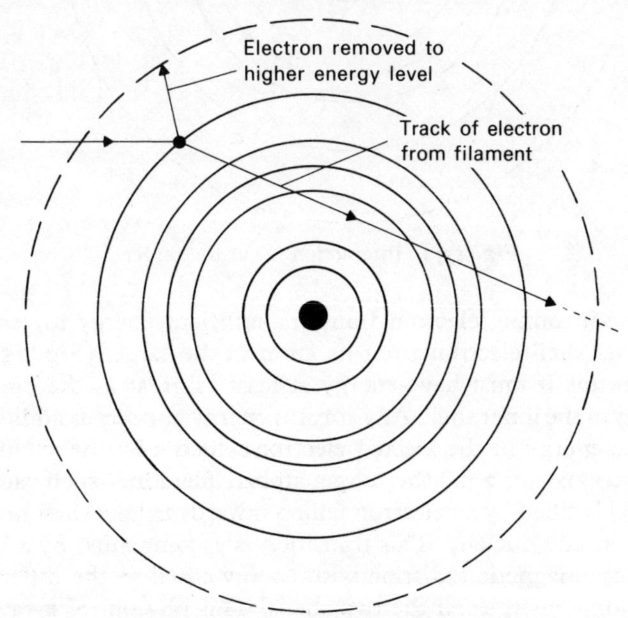

Fig. 15/2. Interaction (1) at the target.

2. The incoming electron transfers enough of its energy to an *outer shell* electron in a target atom to *remove it completely* from its atom; i.e. the atom is ionised. If the electron has enough energy after the collision it may undergo further interactions of types **1** or **2** many times, ultimately causing the target to heat up (Fig. 15/3).

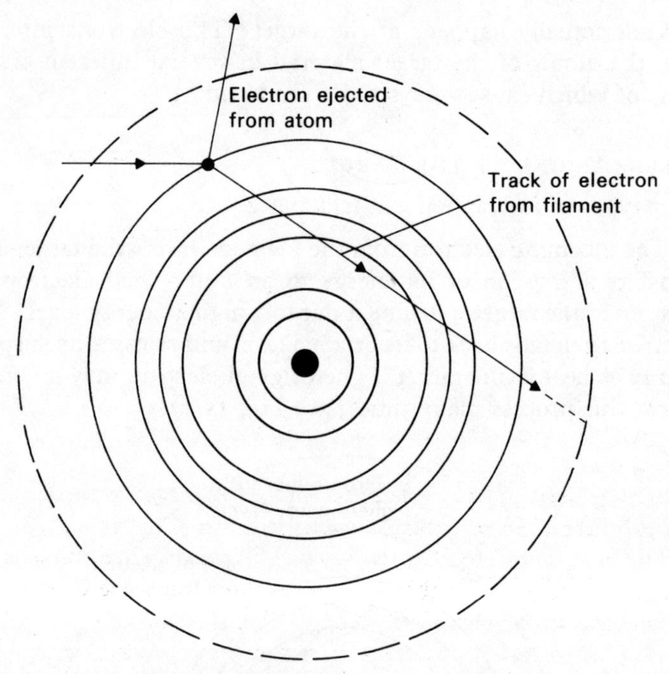

Fig. 15/3. Interaction (2) at the target.

3. The incoming electron transfers sufficient energy to remove an inner shell electron from its atom in the target (Fig. 15/4a). To do this it must have energy at least as great as the binding energy of the inner shell. Any surplus energy appears as additional kinetic energy in the ejected electron which may then undergo interactions **1** or **2** in other target atoms. The inner shell vacancy is quickly filled by an electron falling inwards from a shell further out from the nucleus. This transition is accompanied by a burst of electromagnetic radiation with energy equal to the difference in binding energies of the two shells. This photon (of x-rays) is known as *characteristic x-radiation* because its exact photon energy

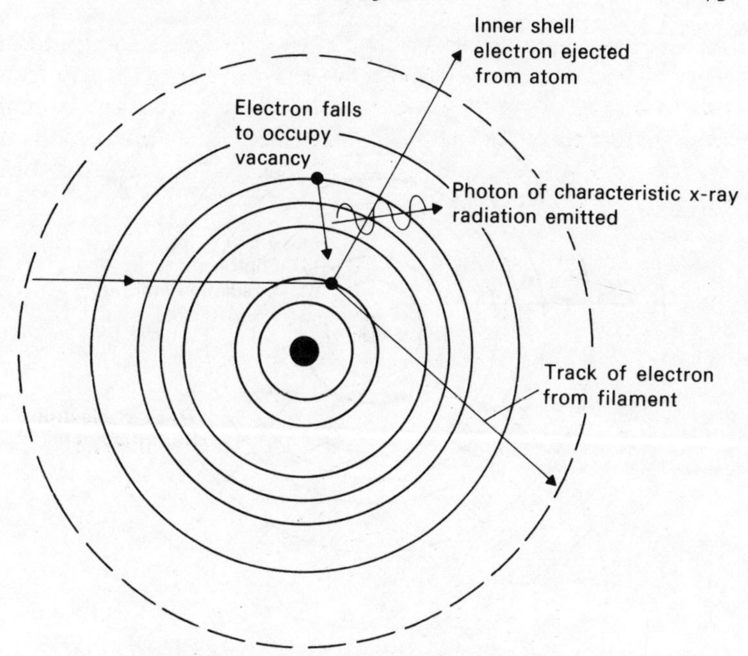

Fig. 15/4. (a) Interaction (3) at the target.

is characteristic of the element of which the target is made. This will produce a characteristic (line) spectrum (Fig. 15/5).

4. The incoming electron passes very close to the nucleus of a target atom (Fig. 15/4b). The attraction causes the electron to deviate and slow down in its course. The slowing down involves a loss of kinetic energy which is transformed into electromagnetic radiation by the process described in Chapter 14. The energy of the radiation will depend on the degree of deceleration which occurs. In an extreme case the electron may actually be brought to rest. Thus the photon energy can be of any value from zero up to a maximum equal to the total kinetic energy of the incoming electron. This gives rise to a *continuous* spectrum of x-radiation and is known as *braking radiation* (Fig. 15/6).

RELATIVE INCIDENCE OF THE INTERACTIONS

Processes **1** and **2** are much more likely to occur than **3** or **4**. In fact *less than 1%* of the energy in a diagnostic x-ray tube is converted into x-rays. At the higher voltages used in radiotherapy

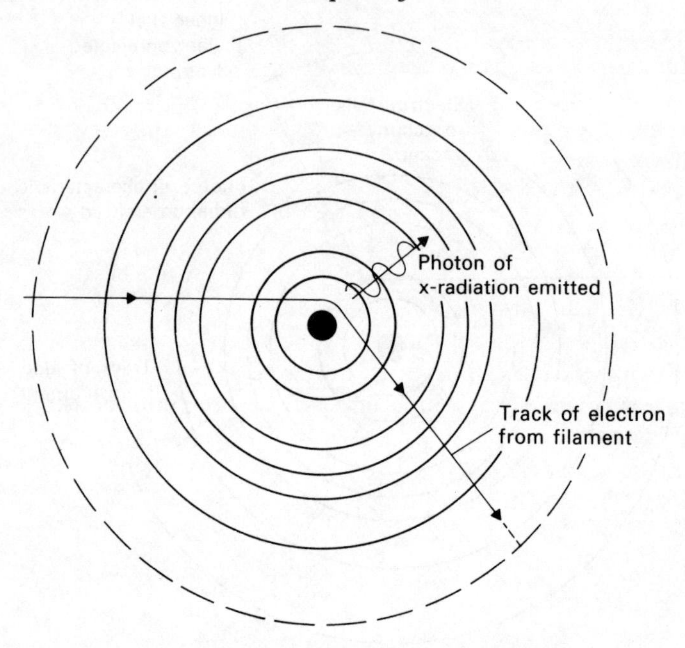

Fig. 15/4. (b) Interaction (4) at the target.

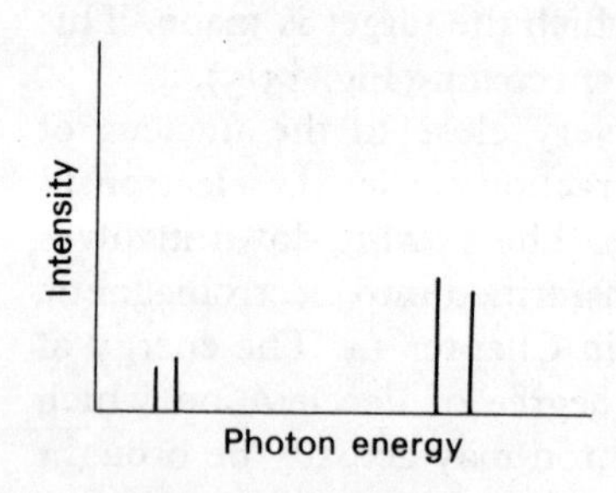

Fig. 15/5. The x-ray line spectrum, showing photons of four different energy values being emitted.

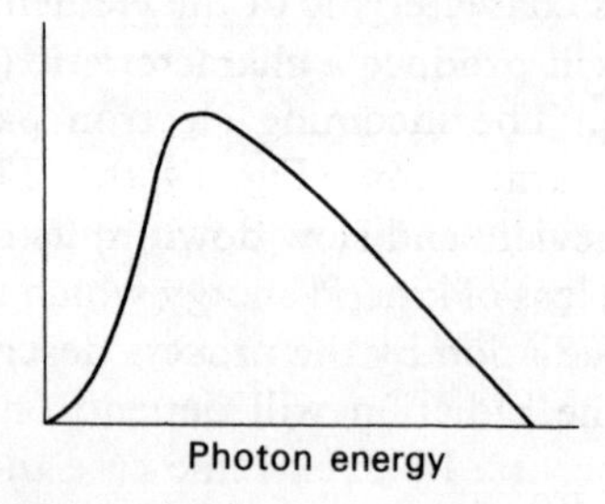

Fig. 15/6. The continuous x-ray spectrum.

the efficiency of x-ray production is higher (up to 30%) and relatively less heat is produced.

Process **3** (characteristic x-ray production) *cannot* take place at all if the x-ray tube voltage is insufficient to give electrons enough energy to remove an inner shell electron from a target atom. With a tungsten target, voltages below about 68 kVp cannot produce tungsten characteristic radiation.

The large amounts of unwanted heat generated in x-ray tubes create enormous problems for the designers of x-ray tubes, as we have seen in Chapter 12.

X-RAY SPECTRUM

As we have discovered, processes **3** and **4** produce line and continuous spectra respectively. However, unlike light, the line x-ray spectrum can never be produced alone in an x-ray tube. If present, it is always superimposed on the continuous spectrum (Fig. 15/7).

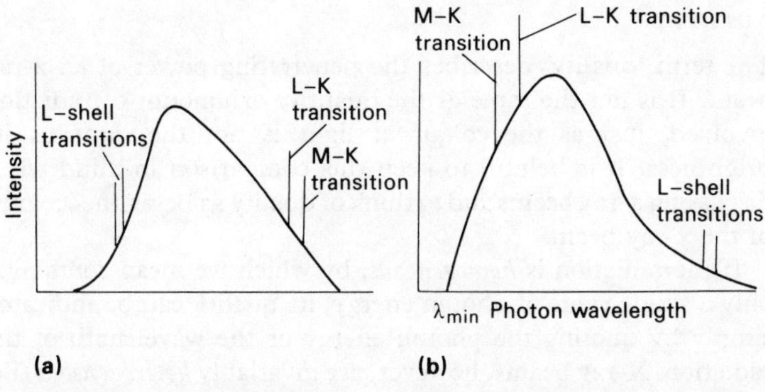

Fig. 15/7. (a) Combining the continuous and line energy spectra.

Fig. 15/7. (b) Combining the continuous and line wavelength spectra.

The two most prominent line spectra from a tungsten target are caused by electrons filling vacancies in the K shell (one line arises from an L–K shell transition, the other from an M–K shell transition). Other lines arise from vacancies being filled in the L shell but the photon energy involved here is too low for it to leave the x-ray tube and it is usually ignored.

MINIMUM PHOTON WAVELENGTH

Another feature of the x-ray spectrum is the presence of an upper photon energy limit (or minimum photon wavelength). For any particular tube voltage there will be a corresponding upper photon energy limit or minimum wavelength. For example at a tube voltage of 86 kVp the maximum photon energy will be 86 keV.

By using the relationship $E = hc/\lambda$ discussed in Chapter 14, we can calculate the value of the minimum wavelength (λ_{min}) if we know the peak value of the tube voltage. When simplified, the relationship can be stated as:

$$\lambda_{min} = 1.24/\text{kVp nanometres.}$$

This relationship is known as the **Duane-Hunt Law**. As the tube voltage is increased, maximum photon energy is increased and minimum wavelength is decreased.

QUALITY AND INTENSITY OF X-RAYS

QUALITY

The term 'quality' describes the penetrating power of an x-ray beam. It is not the same as the quantity or amount of radiation involved, just as the colour of light is not the same as its brightness. It is helpful to keep this comparison in mind when discussing x-ray beams and to think of quality as being the 'colour' of the x-ray beam.

If the radiation is *homogeneous*, by which we mean containing only a single value of photon energy, its quality can be indicated simply by quoting the photon energy or the wavelength of the radiation. X-ray beams, however, are invariably *heterogeneous* (i.e. having a range of different photon energies) and to describe completely the quality of such a beam it would be necessary to give the spectrum of the radiation. In practice in radiography a more useful method is to quote: (a) the generating voltage; (b) the half value layer; and (c) the beam filtration.

We shall be discussing these aspects in detail in Chapter 18.

INTENSITY

This is a measure of the quantity (amount) of radiation energy flowing in unit time. It can be usefully compared with the brightness of light; the 'brightness' of an x-ray beam is its intensity. Intensity is defined as the quantity of energy flowing in unit time through a unit area when measured at right angles to the direction of the beam. Its units are joules/metre^2/second. Fig. 15/8 illustrates the situation referred to in the definition.

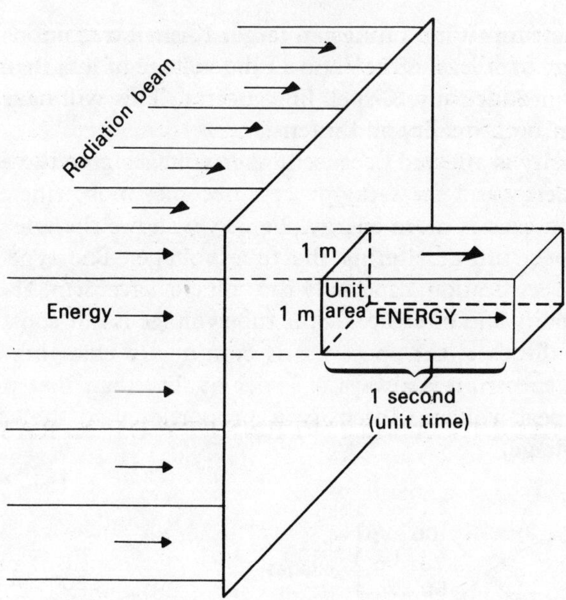

Fig. 15/8. The definition of intensity.

The total intensity of an x-ray beam is represented by the sum of the intensities of the separate components of the beam.

In radiography we usually prefer to measure the quantity of radiation in terms of one of its effects rather than in terms of intensity; e.g. the ionising effect on air, which we call *exposure*, is measured in roentgens, not in the basic units of intensity.

FACTORS AFFECTING QUALITY AND INTENSITY

Four factors are involved here; namely tube voltage (kVp); tube current (mA); tube filtration; and target material in the tube.

Tube voltage. This affects both quality and intensity. It affects quality because the maximum photon energy is determined by the peak tube voltage (as also is the minimum wavelength by the Duane-Hunt Law). The higher the tube voltage (kVp), the more penetrating is the beam.

It affects the line spectrum because if the tube voltage is too low, none of the electrons hitting the target will possess enough energy to cause any electron transitions which would produce a line spectrum. With a tungsten target K-shell transitions require an energy of at least 68 keV. So a tube voltage of less than 68 kVp will not produce any K-shell line spectra. This will have a small effect on both quality and intensity.

Intensity is affected because higher voltages give the electrons more energy and the x-ray process becomes more efficient, thus the beam carries more energy. Fig. 15/9 shows the effect on the x-ray spectrum of altering the tube voltage. The type of high voltage rectification, especially the voltage waveform, also affects both quality and intensity. If the tube voltage is not constant as is usually the case, the spectrum is continually changing and the average spectrum is always of lower quality than that produced by the peak voltage. Intensity is proportional to the square of tube voltage.

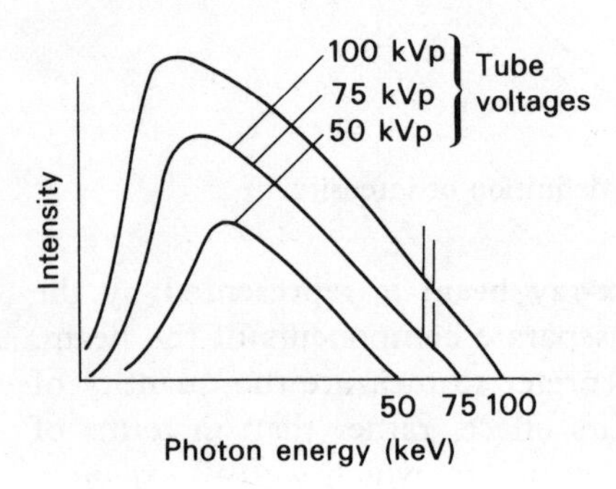

Fig. 15/9. Effect on x-ray spectrum of changing the tube voltage (kVp).

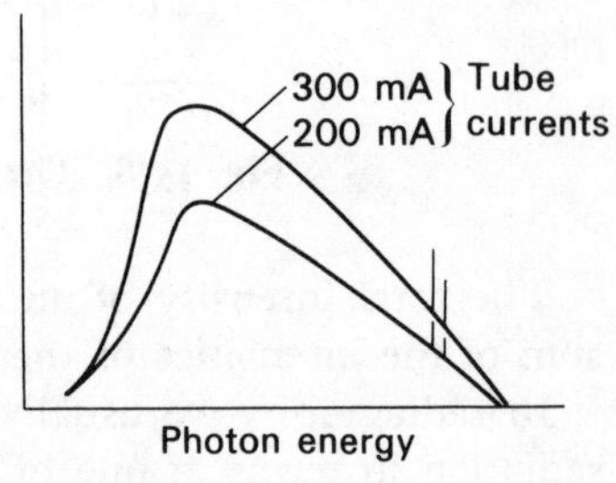

Fig. 15/10. Effect on x-ray spectrum of changing the tube current (mA).

Tube current. The value of tube current (mA) only affects the intensity of the beam; not its quality. It represents the *number* of electrons passing from the filament to the anode. It does not alter the energy of the electrons. The total beam intensity is proportional to the average value of the tube current. Fig. 15/10 shows the effect on the x-ray spectrum of altering the tube current.

Tube filtration. Filters are materials inserted into the x-ray beam to improve the *quality* of the beam. They also reduce the beam's intensity. A filter usually consists of a thin sheet of metal such as aluminium, copper, tin or lead, which has the effect of absorbing most of the low energy (long wavelength) photons and yet transmitting most of the high energy (short wavelength) photons. This is done to reduce unwanted radiation doses to the patient's skin. Fig. 15/11 illustrates the effect on the x-ray spectrum of adding beam filtration (see also Chapter 16).

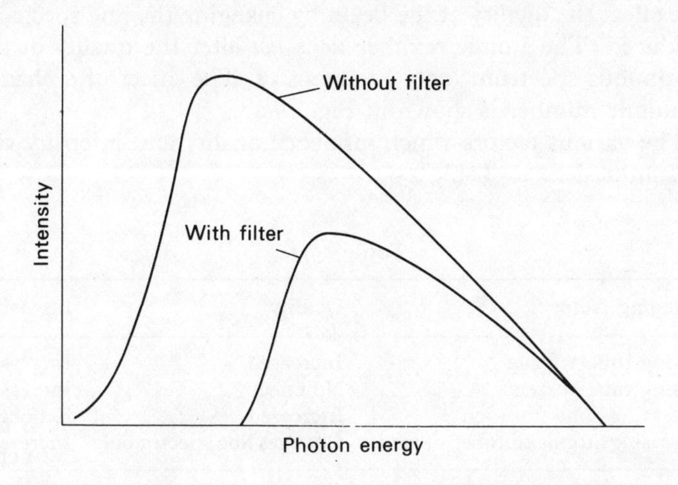

Fig. 15/11. Effect on x-ray spectrum of adding beam filtration.

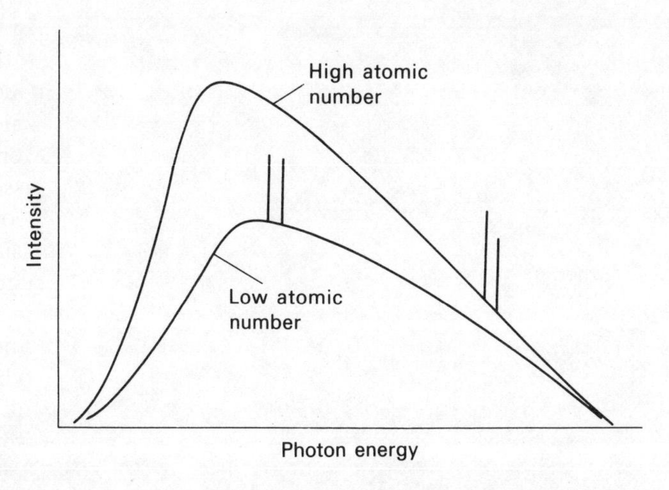

Fig. 15/12. Effect on x-ray spectrum of changing target material.

Target material. The atomic number of the target material affects the intensity of the beam produced. Atoms with high atomic numbers will have a greater decelerating effect on the electrons from the filament because their nuclei have a greater positive electric charge. Thus the x-ray photons produced will have a higher average energy. Also, changing the atomic number will change the photon energy of the characteristic radiation and therefore affect the quality of the beam by changing the line spectrum produced. The atomic number does *not* alter the quality of the continuous spectrum, only its intensity. The effect of a change in atomic number is shown in Fig. 15/12.

The various factors which influence quality and intensity can be summarised in Table 15/1.

Table 15/1.

Changing factor	Quality	Intensity
Raising tube voltage	Increased	Increased
Raising tube current	No effect	Increased
Increasing tube filtration	Increased	Reduced
Increasing atomic number of target	Changes line spectra only	Increased

CHAPTER SUMMARY

1. X-rays are produced when fast moving electrons are slowed down (p. 170).
2. Characteristic radiation is produced when fast moving electrons interact with orbital electrons in the target of the x-ray tube (p. 172).
3. Braking radiation is produced when fast moving electrons interact with the atomic nuclei in the target of the x-ray tube (p. 173).
4. Most of the electrons' energy is converted into unwanted heat in the x-ray tube (p. 173).
5. For each value of peak x-ray tube voltage there is a corresponding maximum x-ray photon energy (p. 175).
6. The quality of an x-ray beam is related to its penetrating power (p. 176).
7. The intensity of an x-ray beam is the quantity of energy flowing through unit area in unit time (p. 176).
8. Tube voltage, tube filtration and tube target material influence both the quality of an x-ray beam and its intensity (p. 177).
9. Tube current affects the intensity but not the quality of an x-ray beam (p. 178).

KEY RELATIONSHIP

$\lambda_{min} = \frac{1.24}{kVp}$ (minimum wavelength) (see p. 176).

Chapter 16
Interaction of x-rays and gamma rays with matter

In this chapter we investigate what happens when a beam of x- or gamma rays is directed at matter. We shall then be in a position to explain how these radiations behave when they interact with living tissues.

A beam of radiation may be transmitted through a medium, e.g. light passes through glass; x-rays pass through wood. However, in each case the intensity of the beam is reduced after it has passed through the matter and we say the beam has been attenuated (made weaker). The loss in intensity may be very small or very great, but there is always some loss of energy from the beam.

Why is this so? What has happened to the radiation energy which is missing?

There are two possibilities:

1. As the beam passes through an object some of its energy is absorbed, i.e. the energy is transferred to matter as in Fig. 16/1. In living tissues this can be a reason for the harmful effects of some radiations.

2. As the beam passes through the matter some of its photons are scattered, i.e. they collide with atomic particles and are forced to change course. They may then emerge but travelling in directions different from the original beam, as in Fig. 16/2. Both of these interactions take place when an x-ray or gamma-ray beam penetrates into matter.

TRANSMISSION OF AN X-RAY BEAM THROUGH A MEDIUM

We saw in Chapter 15 that beams of radiation may be homogeneous (consisting of photons all of the same energy) or heterogeneous (consisting of photons of a whole range of energies).

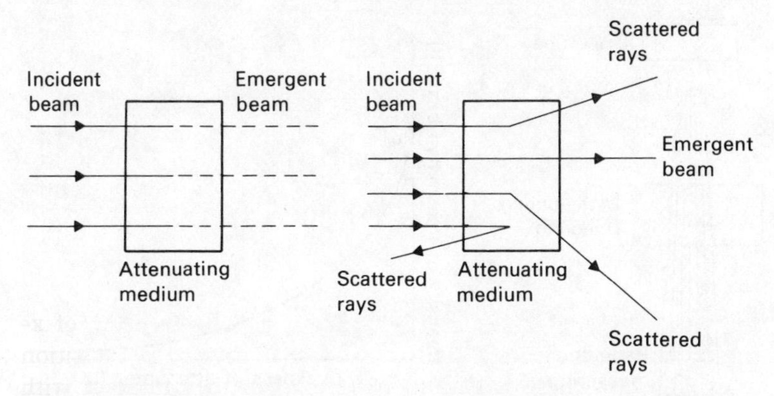

Fig. 16/1. Energy of a beam being absorbed in an attenuating medium.

Fig. 16/2. Energy of a beam being scattered out of the beam.

X-ray beams are always heterogeneous but gamma rays are homogeneous.

It is simpler to consider first what happens when a homogeneous beam passes through a medium and then later to expand the explanation to cover the behaviour of heterogeneous beams.

As we have seen, when a beam of x-rays passes into matter it is attenuated by being absorbed and scattered. How can we investigate the amount of attenuation that takes place? Let us consider an experimental set up (Fig. 16/3) that we could use. It consists of an x-ray source, a radiation-measuring device, and a selection of different thicknesses of a suitable medium such as copper, to insert in the beam. The radiation-measuring device (a dosemeter) will tell us the intensity of radiation transmitted when the beam is attenuated by a known thickness of copper. We take a series of measurements of intensity for increasing thicknesses of copper and plot them on a graph (Fig. 16/4). As commonsense suggests, if thicker attenuating material is inserted, the value of intensity reduces. The graph is the familiar shape that we met with in Chapters 6 and 7 relating to the charging and discharging of capacitors; it is an *exponential decay curve.* It shows that the radiation intensity and thickness of attenuator are related in a very particular way. A mathematical equation can be derived which represents this relationship:

$$I = I_0 e^{-\mu t}$$

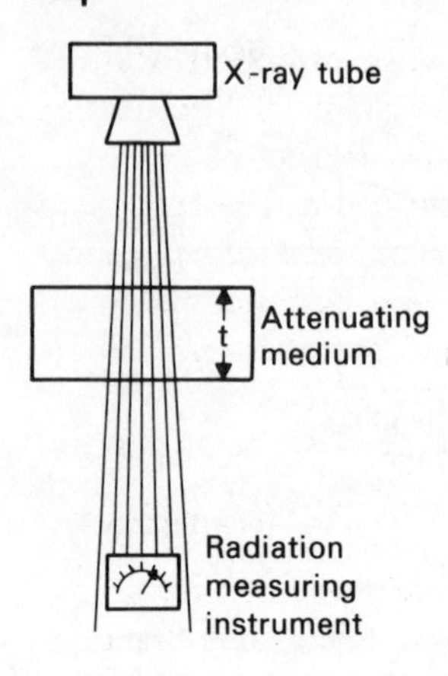

Fig. 16/3. Experimental set-up for measuring attenuation of an x-ray beam.

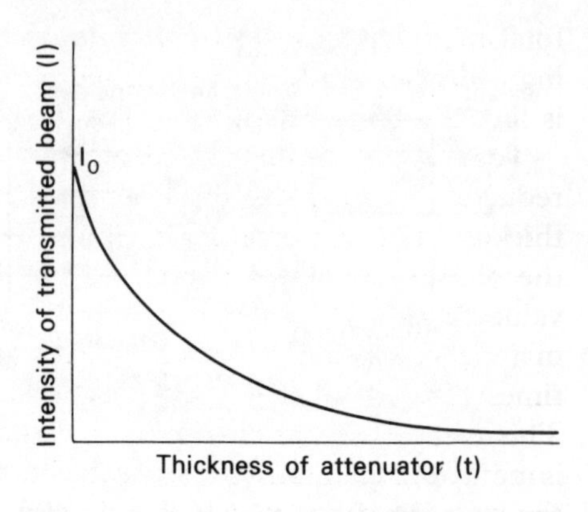

Fig. 16/4. Graph of beam intensity against thickness of attenuator showing exponential relationship.

where:

I is the value of intensity expected when an attenuator of thickness t is inserted in the beam;

I_0 is the value of intensity that is measured when no attenuator is present;

e is the symbol for a number known as the exponential constant. Its value is approximately 2.72;

μ (the Greek letter 'mu') is a quantity which is related to the attenuating properties of the medium through which the beam has passed. It has a different value for different attenuating media. It is known as the *total linear attenuation coefficient*.

t is the thickness of the attenuating material in the beam.

As the exponential decay curve in theory never touches the horizontal axis, no matter how thick the attenuating medium the radiation intensity never reduces to zero.

ATTENUATION COEFFICIENTS

Attenuation coefficients tell us about the effectiveness of different materials as attenuators of radiation. There are several forms of attenuation coefficient employed. The one we have used above, the total linear attenuation coefficient (μ) takes into account the total attenuation caused by all the various absorption and scattering processes involved. It is called a *linear* coefficient because it is linked with the thickness (t) of the attenuating medium.

Total linear attenuation coefficient is defined as the fractional reduction in intensity of a parallel beam of radiation per unit thickness of the attenuating medium. Its value depends partly on the number of atoms per unit volume of the medium, so its value changes even for the same material if the density of the material changes. To avoid this inconvenience its value is sometimes divided by the density (D) of the medium, giving μ/D. This is then used as an alternative attenuation coefficient which is more fundamental in the sense that its value does not alter unless the elements from which the attenuator is made are changed. This coefficient (μ/D) is called the total *mass* attenuation coefficient which we shall abbreviate to TMAC. It is defined as the fractional reduction in intensity in a parallel beam of radiation *per unit mass* of the attenuating medium.

Table 16/1 shows some examples of the values of attenuating coefficients for four different materials when measured using an x-ray beam with a photon energy of 50 keV. It is important to realise that as the photon energy of the beam increases, a greater fraction of the beam will be transmitted through an attenuating medium. The values of μ and TMAC will therefore change if the photon energy of the beam is altered.

Table 16/1. Examples of attenuation coefficients for a 50-keV beam.

Material	Linear attenuation coefficient (μ)	Mass attenuation coefficient (TMAC)
Water	0.2 cm^{-1}	0.2 cm^2/g
Aluminium	1.0 cm^{-1}	0.4 cm^2/g
Calcium	1.5 cm^{-1}	1.0 cm^2/g
Lead	7.0 cm^{-1}	6.0 cm^2/g

Attenuation coefficients help us to predict how much attenuation will occur in a particular set of circumstances. However, they do not help us to explain how or why attenuation occurs. To answer these questions we need to explain the processes which contribute to attenuation.

PROCESSES OF ATTENUATION

Attenuation is caused by absorption and scatter. Four processes need to be described to explain how absorption and scatter occur, and they are:

1. Unmodified scatter (also known as classical or elastic scatter).
2. Photoelectric absorption.
3. Compton scatter (also known as modified scatter).
4. Pair production.

We shall now consider each of these processes in detail.

1. UNMODIFIED SCATTER

This occurs when the energy of photons in the beam is small compared with electron binding energies in the atoms of the attenuating medium, i.e. it occurs with low energy radiation. Fig. 16/5 shows a photon of radiation colliding with an electron in an atom and 'rebounding' away in a different direction. The photon does not have enough energy to release the electron from

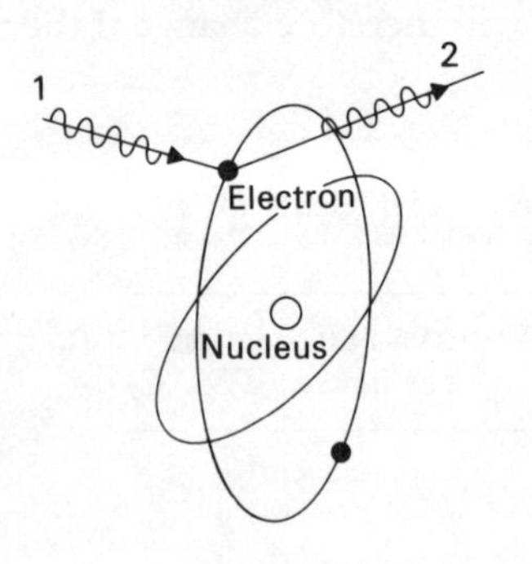

Fig. 16/5. Unmodified scattering of low energy photon. (1) Path of incoming low energy photon; (2) photon deflected by collision with electron (without loss of energy).

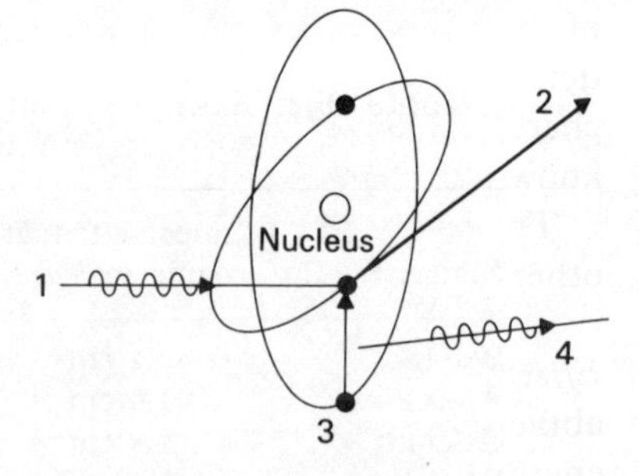

Fig. 16/6. Stages in the process of photoelectric absorption. (1) X-ray photon collides with orbital electron; (2) electron ejected from shell; (3) electron from outer shell fills vacancy; (4) photon of characteristic radiation emitted.

its shell, so none of its energy is transferred to the electron. The photon does not lose energy, the only change being one of direction, hence the term *unmodified* scatter.

This process is not important at the x-ray energies normally used in radiography.

2. PHOTOELECTRIC ABSORPTION

This process occurs when the energy of photons in the beam is equal to or not much greater than the electron binding energies in the atoms of the attenuating medium. The photon transfers *all* its energy to an electron. The electron then has more than the binding energy of its shell and is ejected from the atom (Fig. 16/6). The electron ejected is known as a *photoelectron*. Because the photon has given up all its energy it no longer exists, i.e. true absorption has taken place.

The photoelectric absorption process leaves the atom with an electron vacancy. This is filled by electron transition from an outer shell, accompanied by the emission of a photon of characteristic radiation. The photon of radiation emitted has an energy value which depends on the difference in binding energies of the two electron shells involved in the transition; e.g. an M–K shell transition would produce a photon of higher energy than an L–K transition. The energy also depends on the atomic number of the atom in the attenuating medium because it is that which determines the binding energy values of the electron shells. The characteristic radiation emitted is often itself in the form of x-rays, known as characteristic x-rays.

The photoelectron ejected may continue on to excite and ionise other atoms in the medium before its energy is drained away.

Effect of Photon Energy on Photoelectric Absorption. The probability of photoelectric absorption occurring is greatest when the energy of the incoming photon just equals the binding energy of the electron it interacts with. If the photon energy is below the binding energy the process cannot occur. If the photon energy is much above the binding energy, the chances of the process occurring are reduced. A golfing analogy may be helpful here. When a golfer is trying to hole the ball, he must ensure that he strikes it at the correct speed as well as in the right direction. If the ball is hit too weakly it will not reach the hole and therefore cannot drop in. If the ball is hit too strongly it may pass straight over the hole without dropping in. But if the ball is struck correctly, with just the right force, its chances of falling into the hole are very much greater, assuming, of course, that its direction is correct. However, we should remind ourselves that this is only an analogy and that it is the *energy*, not the speed of photons which must be within certain limits for photoelectric absorption to occur. *All* photons, whatever their energy, travel at the same speed.

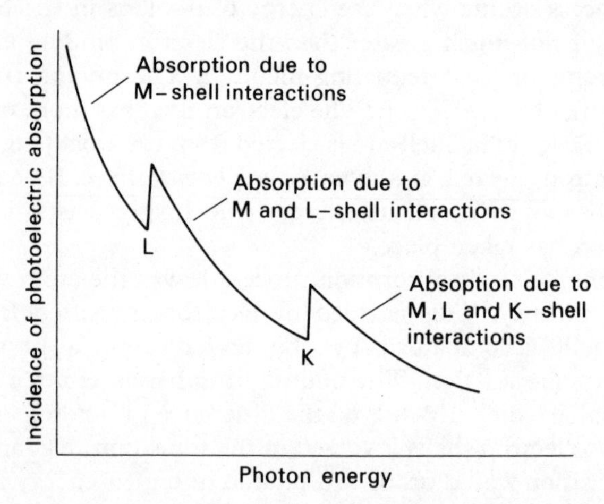

Fig. 16/7. Reduction in photoelectric absorption as photon energy rises.

We can show on a graph (Fig. 16/7) the effect of photon energy on the incidence of photoelectric absorption. As the photon energy rises, the probability of M-shell transition falls until the energy has reached a level high enough to cause L-shell transitions. The probability then rises abruptly. Increasing photon energy further reduces probability again until the binding energy for the K shell is reached. At this point there is another steep rise in probability. Further increases reduce the probability eventually to an insignificant level.

If the incidence of photoelectric absorption is high, more attenuation will take place so the mass attenuation coefficient (MAC) will be correspondingly higher. The total mass attenuation coefficient (TMAC) is the sum of several coefficients each related to a different aspect of attenuation. Thus part of TMAC is due to the attenuation from photoelectric absorption; we shall call it PEMAC. Fig. 16/8 shows how PEMAC changes with photon

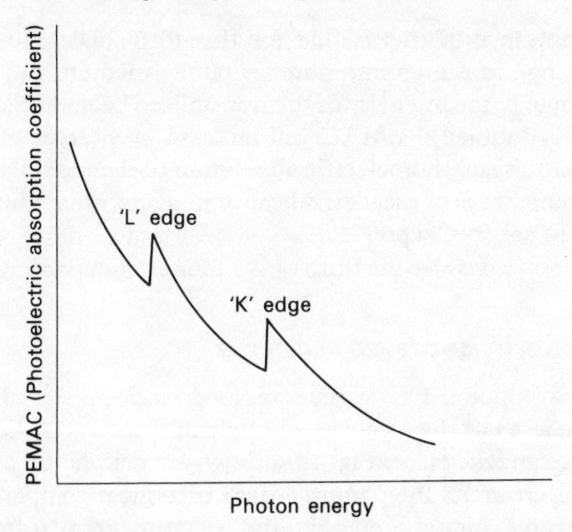

Fig. 16/8. Reduction in PEMAC with rising photon energy.

energy. Clearly this is very similar to Fig. 16/7. The discontinuities, where attenuation is suddenly increased, are called 'absorption edges'. Apart from these anomalies the attenuation due to photoelectric absorption is inversely proportional to the cube of the photon energy (E); i.e.

$$\text{PEMAC} \propto \frac{1}{E^3}$$

Quite small increases in photon energy can therefore produce significant reductions in the amount of photoelectric absorption taking place; for example, if E is doubled, PEMAC will be reduced to one-eighth of its former value.

Effect of Attenuating Medium on Photoelectric Absorption. If the photon energy is kept constant, and a range of different attenuating media is considered, photoelectric absorption occurs to a much greater degree in materials having a high atomic number. It has been found that the attenuation due to photoelectric absorption (PEMAC) is directly proportional to the cube of the atomic number (Z) of the attenuator:

$$\text{PEMAC} \propto Z^3$$

The significance of this is that for this form of attenuation, a slight change in the atomic number of the medium produces a large change in the intensity of the transmitted beam of radiation; (e.g. if Z is doubled, PEMAC will increase by a factor of eight). This sensitivity of photoelectric absorption to the atomic number of the attenuator is of great significance in diagnostic radiography as we shall see in Chapter 17.

Let us now consider the third of the four attenuation processes.

3. COMPTON (MODIFIED) SCATTER

If an x-ray photon has an energy *very much greater* than the binding energy of the electron with which it interacts, Compton scattering can take place. Fig. 16/9 shows such a photon colliding with an electron. As the photon energy is so great compared with the electron's binding energy, the electron recoils from the collision and is ejected at speed from its atom. It is sometimes referred to as a *Compton recoil electron* and it may continue on to excite and ionise other atoms in the attenuator before coming

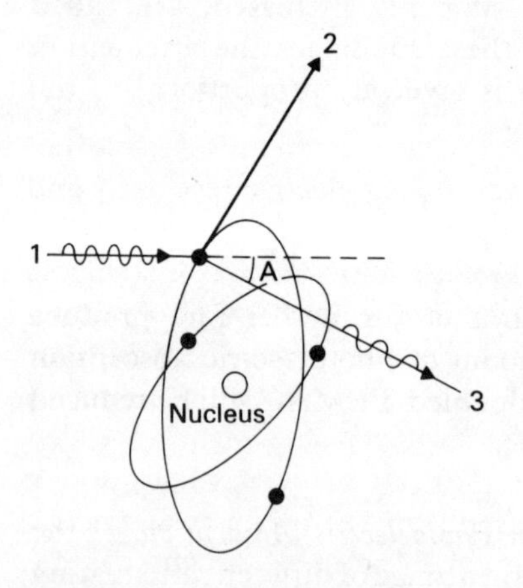

Fig. 16/9. Stages in the process of Compton scattering. (1) X-ray photon collides with orbital electron; (2) electron ejected from atom; (3) photon scattered through angle 'A' continues with reduced energy and longer wavelength.

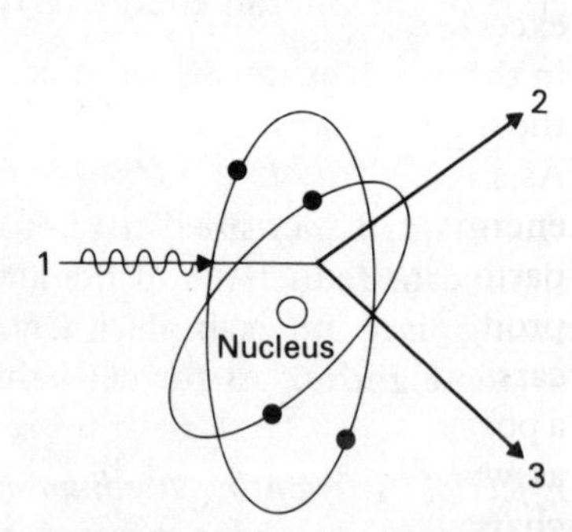

Fig. 16/10. Transformation of photon energy into two particles of matter by pair-production. (1) X-ray photon passes close to nucleus; (2) negative electron (negatron); (3) positive electron (positron).

to rest. The photon has transferred some of its energy to the electron in the form of the electron's kinetic energy so some absorption of energy has occurred. The photon suffers a change of direction as a result of the collision so scattering of the photon has also taken place. The photon's energy is reduced giving it a lower frequency and a longer wavelength.

The loss of energy of the photon depends on the angle (A) through which it is scattered. It is greatest when $A = 180°$; i.e. when the photon is scattered back along its original path. The energy loss does not depend on the photon's original energy, nor on the attenuating medium, but the scattering angle A tends to be smaller for photons with high energies as they are less likely to be deflected off course by collisions with electrons. The incidence of Compton scatter therefore decreases as the photon energy of the beam increases. This affects the mass attenuation coefficient due to the Compton effect (CEMAC), which also decreases as the photon energy is raised. In contrast to PEMAC, at a fixed photon energy the value of CEMAC is the same for all attenuating media.

4. PAIR PRODUCTION

This absorption process can only occur when the photon energy exceeds 1.02 MeV. As the photon passes the nucleus of an atom in the attenuator it experiences the strong electric forces around the nucleus caused by the positive charges on the protons there. As a result, the photon undergoes a dramatic change of state: its energy is transformed into matter in the form of two minute particles, as shown in Fig. 16/10, hence the process is called pair production. The particles are electrons; one, called a negatron, carries a negative charge as do 'normal' electrons, the other, called a positron, carries an equal and opposite positive charge. A positive as well as negative charge is produced so that the total electric charge remains zero, since the original photon carried no charge.

At least 1.02 MeV is needed to produce this effect because this represents the energy equivalent of the masses of two electrons (derived from $E = mc^2$). Any energy the photon possesses in excess of 1.02 MeV is converted into kinetic energy of the two particles, both of which can excite and ionise other atoms in the attenuating medium. When, however, the positron eventually comes to rest, it combines with a 'normal' electron and the two are transformed from matter back into radiation energy. The radiation produced in this way is called annihilation radiation since it results from the annihilation of two particles. It appears in the form of two 0.51 MeV photons.

Pair production is a process of true absorption as all of the energy of the original photon is transformed.

The mass attenuation coefficient due to pair production (PPMAC) is zero at photon energies below 1.02 MeV and increases as energy increases above that value (Fig. 16/11). For a fixed photon energy PPMAC is directly proportional to the atomic number (Z) of the attenuating material, i.e.

$$\text{PPMAC} \propto Z.$$

It is important to note that pair production *never* occurs with the x-ray energies used in diagnostic radiography.

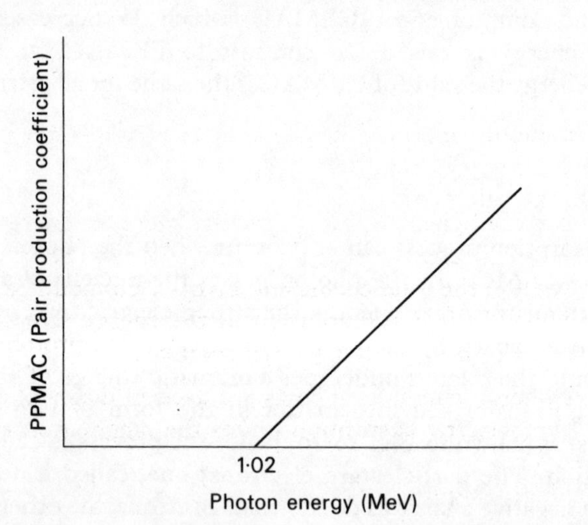

Fig. 16/11. Increasing attenuation due to pair-production at photon energies above 1.02 MeV.

RELATIVE IMPORTANCE OF THE ATTENUATION PROCESSES

It is useful to be able to illustrate how the relative importance of the processes of attenuation we have described changes for different radiation energies. Fig. 16/12 is a graph of mass

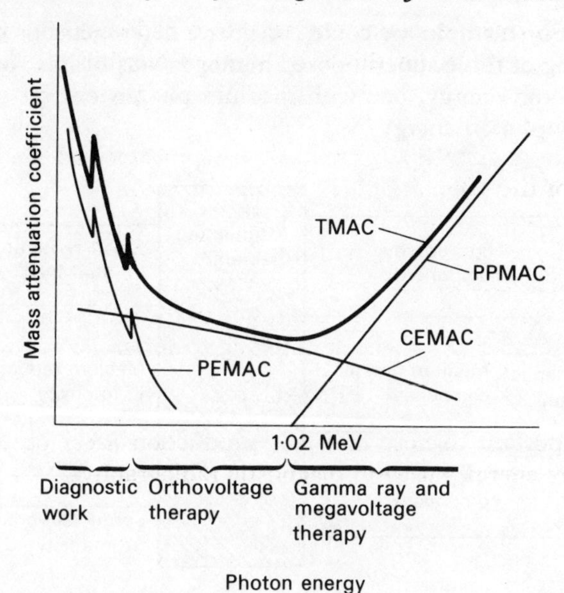

Fig. 16/12. Relative incidence of the attenuation processes over a range of photon energies.

attenuation coefficients plotted against photon energy. The individual coefficients (PEMAC, CEMAC, and PPMAC) are shown as well as the total coefficient TMAC. Unmodified scatter is not included as it is unimportant in the energy range we are considering. From the graph we can see that:

1. At low photon energies, such as are used in diagnostic radiography, photoelectric absorption makes the biggest contribution to attenuation;

2. At medium energies, such as are used in orthovoltage radiotherapy, Compton scatter effect makes the biggest contribution.

3. At the high photon energies employed in megavoltage and gamma-ray radiotherapy, pair production makes a major contribution.

ATTENUATION OF HETEROGENEOUS BEAMS

Up to now we have only considered the transmission of homogeneous beams through a medium. However, we can think of heterogeneous beams as a combination of many homogeneous beams. For example, we could consider a heterogeneous beam as consisting of three superimposed homogeneous beams; one with high photon energy, one with medium photon energy and one with low photon energy.

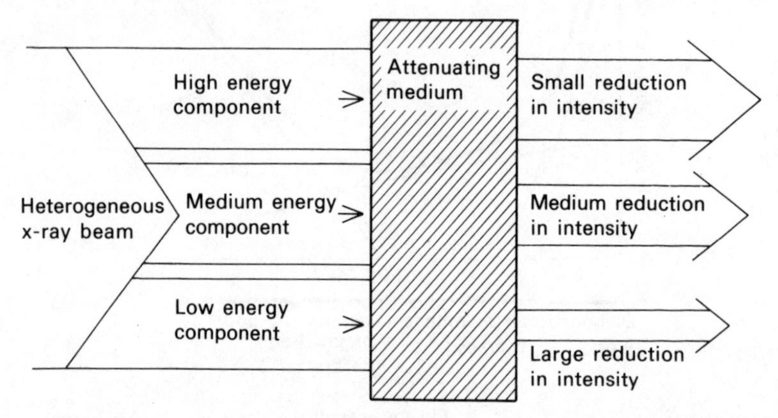

Fig. 16/13. Modification of a heterogeneous beam by attenuation.

Fig. 16/13 shows such a beam being transmitted through an attenuating medium. We can see that the high energy component of the beam is attenuated to a lesser degree than the low energy component. The transmitted (emergent) beam, although reduced in intensity, contains a greater proportion of high energy photons than did the original beam before attenuation. Most of the low energy photons are absorbed or scattered from the beam. With sufficient attenuation we can transform a beam which contains a complete range of photon energies into a beam which is almost homogeneous, containing mainly high energy photons.

A similar effect can be seen if a beam of white light is passed through a coloured sheet of glass. The white light is heterogeneous since it contains many different photon energies, but after passing through the glass only a few specific photon energies remain and the emergent light beam is homogeneous (monochromatic).

The effect of attenuation we have been describing is termed *filtration*, and the attenuators used for the purpose of modifying the beam in this way are known as filters. It may be useful to consider an analogy from the world of athletics to help with our understanding of the idea of filtration. In athletics competitions, preliminary rounds known as 'heats' are used to 'filter out' the weaker competitors, so that by the time the final round is reached the quality of those taking part is high. We can think of filters of radiation as having the same sort of effect on x-ray photons as athletic heats do on the competitors.

FILTRATION

Any medium through which a heterogeneous beam of radiation passes modifies the spectrum of the beam. Fig. 16/14 illustrates the spectrum of a beam of x-rays before and after passing through

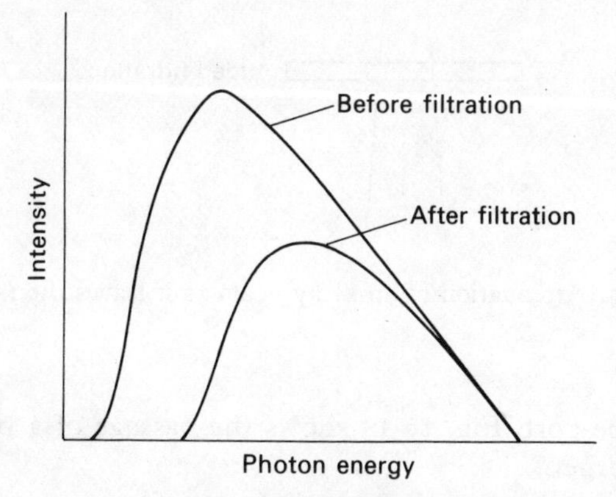

Fig. 16/14. The effect of filtration on the spectrum of a heterogeneous x-ray beam.

a filter. It shows that although the intensity of the beam is reduced, its quality is improved because the low energy component is virtually removed from the beam. Very low energy photons are not needed in the beam because in radiotherapy they do not penetrate to the deeper tissues of the body which are often being treated, and in diagnostic work they do not reach the x-ray film and therefore do not contribute to the formation of the radiographic image. In both branches of radiography low energy photons are positively harmful as they produce undesirable doses of radiation to the patient's skin.

For biological safety reasons, therefore, beam filtration is an important feature of equipment design. Some degree of filtration is inevitable because an x-ray beam has to pass through parts of the x-ray tube itself before emerging from the tube port; e.g. (a) the target; (b) the glass envelope; (c) the insulating oil; and

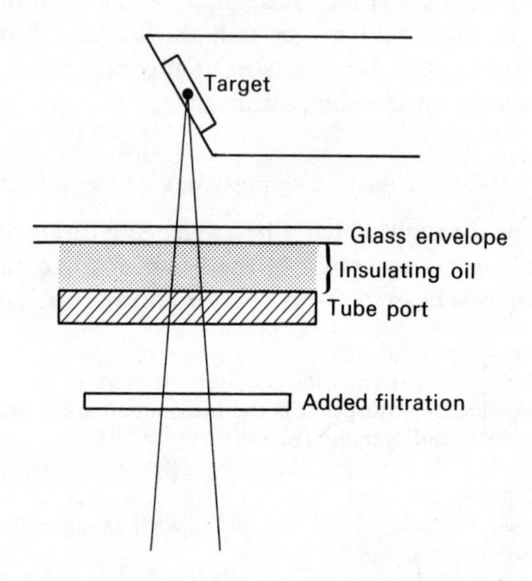

Fig. 16/15. Attenuation of an x-ray beam as it leaves the x-ray tube.

(d) the tube port. Fig. 16/15 shows the passage of a beam from the tube target.

In diagnostic x-ray tubes, this inherent filtration is equivalent to the filtration due to a layer of aluminium 1 mm thick; i.e. the inherent filtration is equivalent to '1 mm Al'. To produce sufficient effect on the beam, extra filtration is employed, known as *added* filtration, in the form of thin sheets of metal such as aluminium, copper or tin, to bring the total filtration up to 2 or 2.5 mm Al in diagnostic radiography and even more in radiotherapy. The added filters are attached to the tube casing over the port.

Filter materials have to be chosen with care to ensure that the beam is modified in the way desired, and to ensure that the added filtration is not so thin as to be impractical to make accurately. Aluminium is the most common filter used in diagnostic radiography, while copper, tin and lead may be used in radiotherapy. Recommended levels of filtration are set out in the 'Code of Practice' and we shall be outlining these in Chapter 20.

In the next chapter, we shall relate the processes of interaction we have been describing to the action of x- and gamma ray beams on living tissues.

CHAPTER SUMMARY

1. An x-ray beam is attenuated when transmitted through a medium (p. 182).
2. Attenuation is due to absorption and scattering of radiation (p. 182).
3. The total linear attenuation coefficient is the fractional reduction in intensity of a parallel beam of radiation per unit thickness of the attenuating medium (p. 184).
4. The total mass attenuation coefficient is the fractional reduction in intensity of a parallel beam of radiation per unit mass of the attenuating medium (p. 184).
5. Photoelectric absorption is the main attenuation process in diagnostic radiography (p. 187).
6. Compton scattering is the main attenuation process in radiotherapy (p. 190).
7. Pair production only occurs when photon energies exceed 1.02 MeV (p. 191).
8. The quality of a heterogeneous beam is improved by transmission through an attenuating filter (p. 194).
9. Beam filtration is employed in diagnostic radiography for biological safety reasons (p. 195).

SOME KEY RELATIONSHIPS

$I = I_0 e^{-\mu t}$ (total linear attenuation coefficient) (p. 183).

Mass attenuation coefficient $= \dfrac{\mu}{D}$ (p. 185).

PEMAC $\propto \dfrac{1}{E^3}$ (p. 189).

PEMAC $\propto Z^3$ (p. 189).

PPMAC $\propto Z$ (p. 192).

Chapter 17
X-ray and gamma ray interaction with tissues

Having explained the processes of attenuation which occur when a radiation beam interacts with matter, we can investigate the particular case of a beam passing through living body tissues. We shall then be better able to understand the treatment of malignant tissues by radiation and how the beam which is transmitted through the patient may be used to produce an image on a radiograph or television screen.

TRANSMISSION OF X-RAY BEAMS THROUGH BODY TISSUES

We have seen that when a beam of x- or gamma rays interacts with matter, four possible attenuation processes may occur: namely (a) unmodified scatter (though not important in radiography); (b) photoelectric absorption; (c) Compton (modified) scatter; and (d) pair production (at energies above 1.02 MeV).

As living tissue is a form of matter, consisting of the same kinds of atoms and fundamental particles, it is reasonable to suppose that these same four processes occur when body tissues are exposed to radiation. We saw in Chapter 16 that different materials produce different degrees of attenuation, according to the values of their total attenuation coefficients. Different types of body tissue have different total attenuation coefficients, too, and therefore attenuate x- or gamma-ray beams differently. We know that the value of the total attenuation coefficient of a substance depends largely on its atomic number (Z). Let us therefore examine the atomic numbers of some types of body tissue. Tissues are composed of many different chemical elements, each with its own specific value of atomic number, but we can make an estimate of the average or effective atomic number of a particular tissue, taking into account the individual elements present and their relative abundance in the tissue; for example:

1. Bone has an effective atomic number of about 14 because bone contains calcium ($Z = 20$) and phosphorus ($Z = 15$) as well as lighter elements;
2. Soft tissues (muscle, fat, etc.) have an effective atomic number of about 7 because the heavier elements are less abundant. If we assume, then, that for bone, $Z = 14$; and for soft tissue, $Z = 7$; we can see how x- or gamma-ray beams would be affected by interaction with each of these tissues.

The photoelectric absorption process is very dependent on the atomic number of the absorbing medium. We saw in Chapter 16 that its attenuation coefficient, PEMAC, is proportional to Z^3. We may infer, therefore, that at a particular beam energy, PEMAC for bone is eight times (2^3) higher than PEMAC for soft tissue since the atomic number of bone is twice that of soft tissue. So bone causes a lot more photoelectric absorption than does soft tissue.

The Compton scattering process is relatively independent of the atomic number of the scattering medium. In this case we may infer that bone and soft tissue cause roughly the same amount of Compton scattering of an x- or gamma-ray beam.

The pair production process depends on atomic number, but not to the same extent as in the case of photoelectric absorption. In fact, PPMAC is directly proportional to Z, and PPMAC for bone is roughly twice PPMAC for soft tissue at a particular beam energy.

Let us now see what this means in practice.

DIAGNOSTIC RADIOGRAPHY

If we use a beam energy for which photoelectric absorption is the main process of attenuation (i.e. the range of energies used in diagnostic radiography), soft tissues will transmit a much greater part of the beam than will bone. In Fig. 17/1, the parts of the x-ray film receiving radiation which has passed through soft tissue will be exposed to a far greater extent than the parts which receive the radiation transmitted through bone. When processed, the film will show dark areas where the beam passed through soft tissues, and light areas where the beam passed through bone.

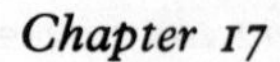

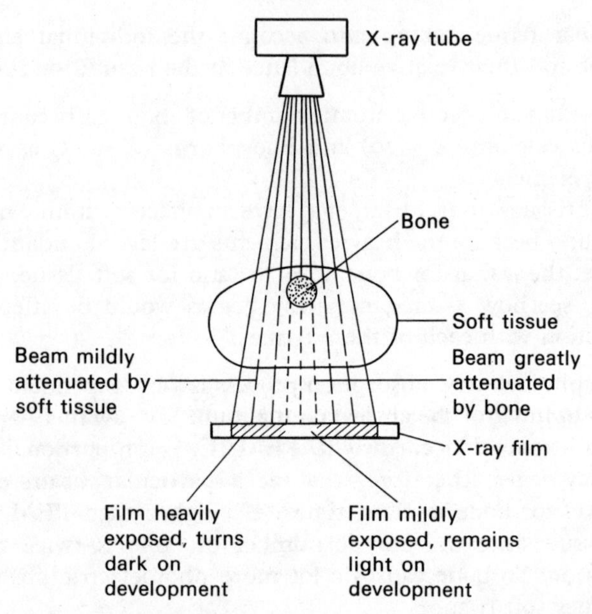

Fig. 17/1. Formation of diagnostic image on x-ray film due to differential absorption in tissues.

Because of this contrast of tones a recognisable image of the bone will be produced.

Of course, the thickness of the tissue through which the beam passes also affects the amount of transmission. A thick layer of soft tissue may produce the same attenuation of the beam as a thin layer of bone. However, as long as we use beam energies for which photoelectric absorption is the main cause of attenuation, then on our radiographs we can expect to be able to differentiate between different types and different thicknesses of tissue. This is why photon energies of 25 to 150 keV and tube voltages of 25 to 150 kVp are used in diagnostic work.

When we modify exposure factors in diagnostic radiography, by increasing the tube kVp, we lower the incidence of photo-electric absorption and the difference in transmission through bone and soft tissue is reduced. There is then less contrast between the tones on the radiographic image. This is a very useful method available to radiographers to control the contrast of the image. Raising the kVp lowers the image contrast, and conversely lowering the kVp increases the image contrast. When very tiny tissue differences are being examined, as in radiography of the female breast (mammography), tube kVp is reduced to a minimum to enhance the very small differences in attenuation so they may be recorded on the x-ray film. Modification of exposure factors *without* changing the kVp does *not* cause alterations in image contrast, because in this case the photon energy of the beam is not affected and there is therefore no change in the incidence of photoelectric absorption.

THERAPEUTIC RADIOGRAPHY

If we use a beam energy for which Compton scattering is the main attenuation process (i.e. the range of energies used in orthovoltage and external beam therapy of deep-seated lesions), both bone and the soft tissues will absorb similar amounts of radiation energy. If lower energies were used, bone would absorb much higher doses of radiation than soft tissue and the treatment of malignant tumours would be more hazardous. We shall see in the next section of this chapter that the scattered radiation caused by the Compton process is of great value in the contribution it makes to the overall effect on a tumour.

EFFECTS OF SCATTERED RADIATION ON BODY TISSUES

As a beam of x- or gamma radiation passes through tissues some of the photons are scattered due to the Compton process. This scattered radiation makes an important contribution to the energy or dose absorbed by the tissue. The total dose received by the tissues is the sum of the dose caused by absorption of the primary beam, and the dose caused by absorption of the scattered radiation.

Any radiation doses to living tissues can be damaging, whether from scattered radiation or the primary beam. Therefore in radiotherapy it is vital to try to restrict the effects of scattered radiation to the tissues being treated, while in diagnostic radiography every effort must be made to ensure that the tissues receive only the very minimum dose consistent with the production of a useful radiograph. To control scatter effectively we need to know what determines how much scatter will be generated.

Because scattered radiation arises from the Compton effect, the amount of scatter generated depends on how much Compton scattering takes place. The volume of tissue irradiated by the primary beam also has an important influence on the quantity of scatter produced. Clearly, more scattered radiation will be generated if a large volume of tissue is irradiated than if a small volume of tissue were irradiated. In practice, therefore, the radiation beam field size (i.e. the diaphragm, cone or collimator setting) can be used to control the quantity of scattered radiation produced.

DIAGNOSTIC RADIOGRAPHY

Scatter has three important effects:

1. It contributes to the radiation dose received by the patient.
2. It contributes to the radiation dose received by the staff.
3. It degrades the quality of the radiographic image.

Patient dose. The amount of scatter generated inside the patient may be minimised by beam collimation ('coning down') because this reduces the volume of tissue irradiated by the primary beam (Fig. 17/2). When the energy of the primary beam is low, i.e. at low kVp values, the scattered radiation is also of low energy and

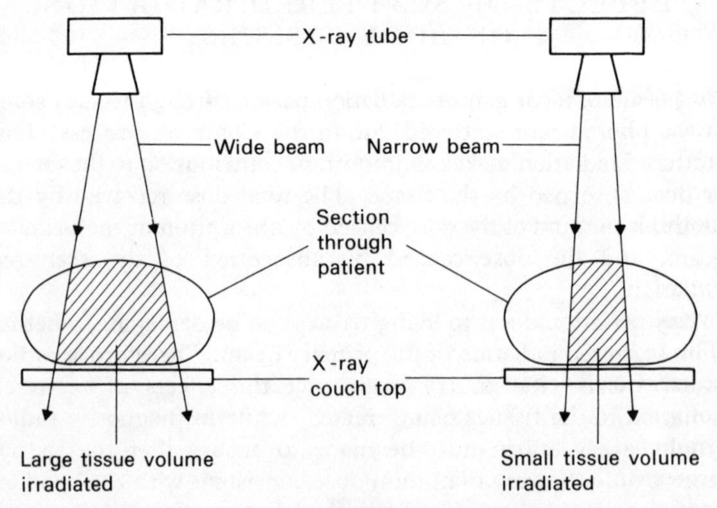

Fig. 17/2. The effect of beam collimation reducing the volume of tissue irradiated.

is likely to be absorbed in the patient in regions close to the area under examination. If the primary beam energy is raised, by increasing the tube kVp, the scatter becomes more penetrating. More scatter will then escape from the patient, and more will produce radiation doses in parts of the patient remote from the original exposure. Fig. 17/3 illustrates how the dose received by the gonads during chest radiography is affected by beam energy.

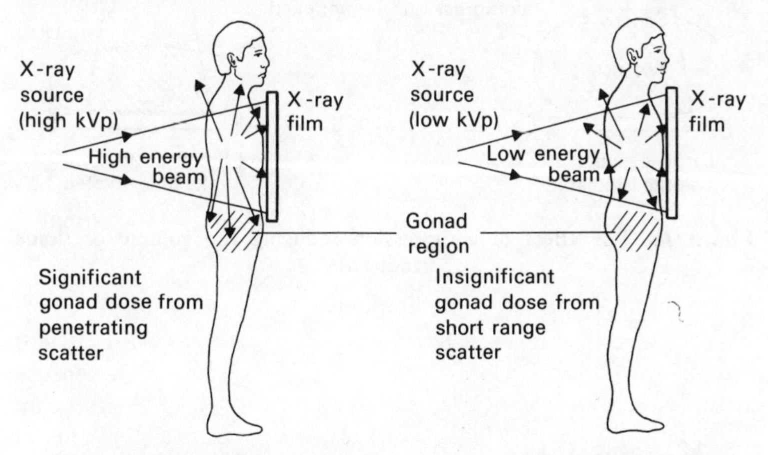

Fig. 17/3. The effect of x-ray beam energy (kVp) on gonad dose.

Staff dose. With low kVp exposures little scattered radiation leaves the deep tissues of the patient (it is mostly absorbed inside the patient). With high kVp exposures more of the scatter escapes from the patient and becomes a hazard to the members of staff in the vicinity. Beam collimation is again an effective method of reducing the danger, but ray-proof barriers and protective clothing offer extra protection.

Image quality. Scattered radiation reaching the x-ray film causes overall fogging of the image, reducing the information it records. The radiographer can overcome this by reducing the amount of scatter being produced by beam collimation and in some cases by compression of the patient to reduce the volume of tissue irradiated, as shown in Fig. 17/4. The radiographer can also protect the film from scattered radiation by the use of secondary radiation grids (Fig. 17/5) and back-scatter protection underneath the film (Fig. 17/6).

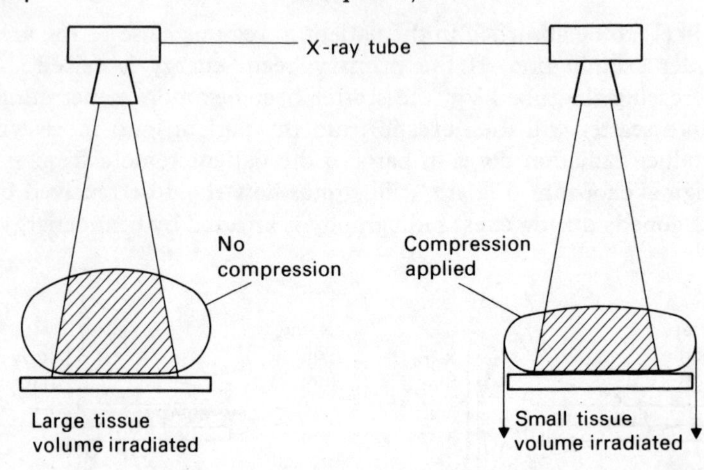

Fig. 17/4. The effect of compression reducing the volume of tissue irradiated.

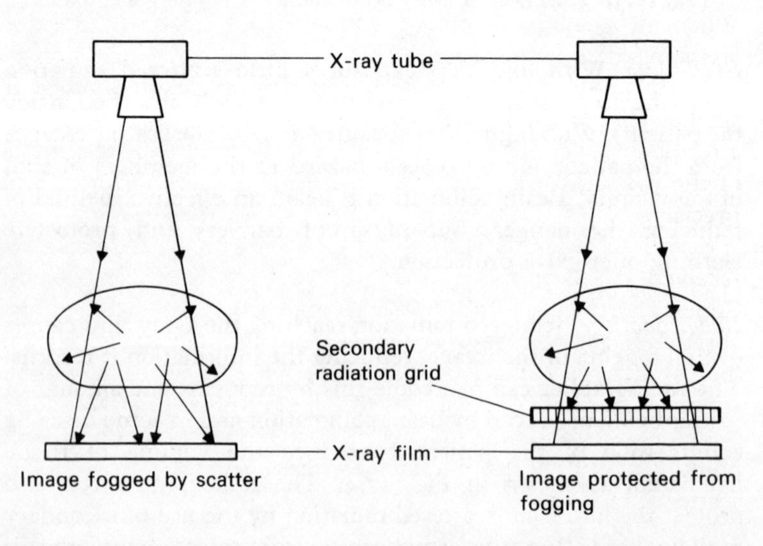

Fig. 17/5. The action of a grid in protecting the radiographic image from fogging by scatter from patient.

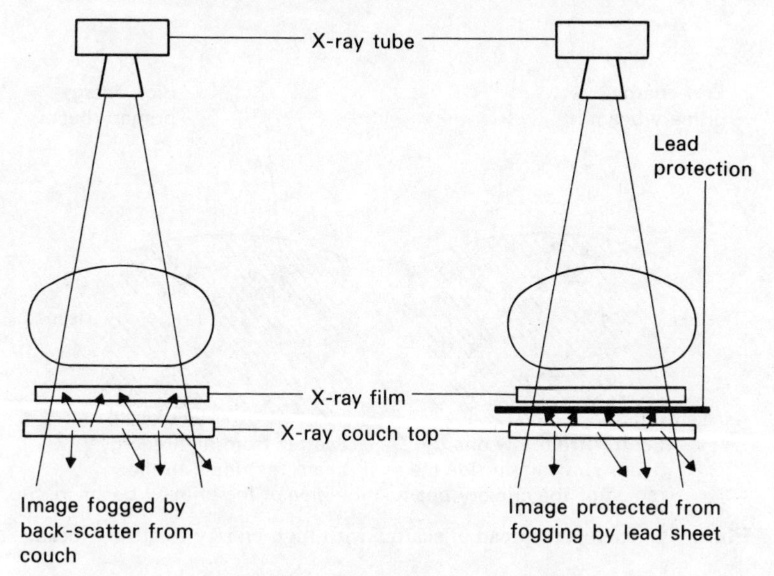

Fig. 17/6. The action of lead backing layer in protecting the image from fogging by back-scatter from the couch top.

THERAPEUTIC RADIOGRAPHY

When high energy beams are employed for treatment there is less spreading of the scattered rays away from the direction of the primary beam because the scattering angle of photons is smaller than it would be for low energy beams (Fig. 17/7). As a result there is less dose to the tissues lying outside the limits of the primary beam, and the treatment is more localised and precise.

Scattered radiation makes a greater contribution to the dose received by deep tissues than to that by superficial tissues. Superficial lesions may not absorb enough dose because of the absence of sufficient scattered radiation. In some cases additional 'bolus' material is placed between the patient and the radiation source to generate extra scatter for this purpose.

Before we leave the topic of scattered radiation it is worth pointing out two of the practical consequences of the properties of scattered radiation:

1. The photon energy of scattered radiation is less than the photon energy of the primary beam from which it is derived.

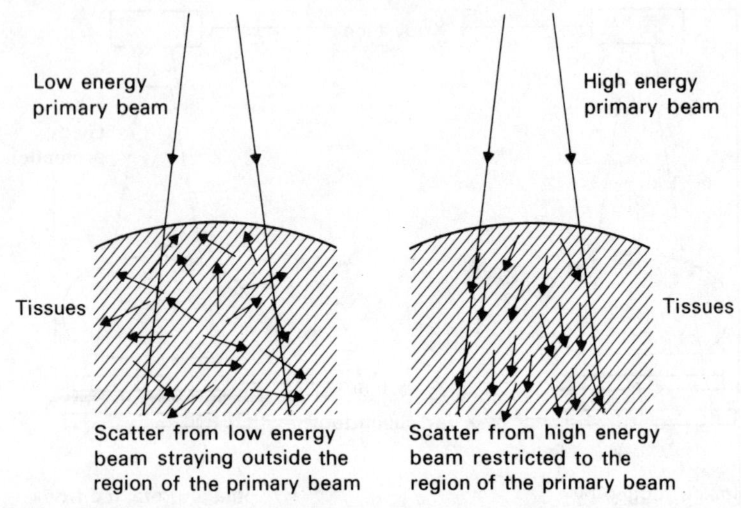

Fig. 17/7. Reduced spread of scatter with high-energy treatment beams.

If we are exposed to scattered radiation we are likely to absorb it because it is of low energy and it therefore constitutes a serious hazard to health.

2. It is not unknown for radiographers to imagine that scattered radiation wanders around the x-ray or treatment room, or is released from the body of a patient, for some time after the primary beam has been switched off. Consideration of the speed at which electromagnetic rays travel should tell us that this is not so. Scattered radiation is only present while the primary beam is present. In that sense we can control scatter very effectively and instantaneously by turning off the primary beam.

In our next chapter we examine some of the methods available for measuring x- and gamma radiation and we take a closer look at the concept of radiation dose.

CHAPTER SUMMARY

1. The same processes of attenuation occur when x-rays interact with living tissues as when they interact with inorganic matter (p. 198).
2. Diagnostic x-rays are able to differentiate different types of body tissue according to their effective atomic numbers (p. 199).
3. Increasing the tube kilovoltage reduces the image contrast between tissues on a radiograph (p. 200).
4. High energy therapeutic x-rays are attenuated equally by the various body tissues (p. 201).
5. In diagnostic radiography, scattered radiation adds to the patient and staff dose and degrades the quality of the radiographic image (p. 202).
6. In therapeutic radiography, scattered radiation contributes to the useful dose to the lesion under treatment (p. 205).
7. The photon energy of scattered radiation is less than that of the primary beam (p. 205).
8. Scattered radiation is only present when the primary beam is present (p. 206).

Chapter 18
X-ray measurements (dosimetry)

ABSORBED DOSE

When matter is irradiated with x- or gamma rays it is important to know how much energy is transferred from the radiation to the material exposed. This is particularly so where the material is living tissue because the effect of radiation on tissue is closely related to the amount of energy it absorbs. The quantity of energy transferred from radiation is known as the *absorbed dose* and is defined as the energy absorbed per unit mass of the medium.

UNITS OF ABSORBED DOSE

The SI unit of absorbed dose is the gray (Gy) which represents one joule of energy absorbed per kilogram of material. There is another unit of absorbed dose in common use called the rad (an abbreviation of Radiation Absorbed Dose). The rad represents 1/100 joule of energy absorbed per kilogram of material. The rad is still (in 1979) the most frequently used unit of absorbed dose, but it is gradually being replaced by the gray. There is a simple relationship between the two units:

$$1 \text{ Gy} = 100 \text{ rad},$$
$$\text{and} \quad 1 \text{ rad} = 1/100 \text{ Gy}.$$

There are also subunits such as the milligray (mGy) where 1 mGy = 1/1000 Gy, and the millirad (mrad) where 1 mrad = 1/1000 rad.

The value of absorbed dose depends on both the photon energy of the beam and on the type of absorbing medium. A high energy beam produces *less* absorbed dose than does a low energy beam of the same intensity because more of its photons are transmitted without absorption. A material having a high effective atomic number receives a higher absorbed dose than a material of low atomic number exposed to the same beam because more absorption interactions take place.

Dose can be expressed in terms of a total (cumulative) value, measured in rads or grays. However, it is sometimes more convenient to express it in terms of doserate, which is the dose absorbed per unit time. This is then quoted in units such as grays/second, milligrays/hour, rads/second, rads/minute or millirads/hour, etc. Total dose and doserate are related since:

$$\text{Total dose} = \text{doserate} \times \text{time}$$

where time represents the length of time over which the total dose is measured. For example:

1. A doserate of 10 millirads/hour continuing for 30 minutes would give:

$$\text{Total dose} = 10 \times \frac{30}{60}$$

$$= 5 \text{ millirads.}$$

2. If a total dose of 45 grays is required in a time of 15 minutes, the doserate required is given by:

$$\text{Doserate} = \frac{\text{total dose}}{\text{time}}$$

$$= \frac{45}{15}$$

$$= 3 \text{ grays/minute.}$$

We shall see in Chapter 20 that there is another method of expressing radiation dose which is more closely related to the effects of radiation of living tissues.

MEASUREMENT OF DOSE

To measure directly the quantity of energy a medium absorbs due to exposure to radiation is technically a very difficult task; it is not feasible to carry it out in the clinical situation in hospital.

In practice, therefore, the value of absorbed dose is determined indirectly by using one of the effects of radiation which can be more easily measured. Suitable effects include:

1. Ionisation of air.
2. Fogging of photographic emulsion.
3. Thermoluminescence.
4. Fluorescence.

Measuring instruments which employ any of these effects as a means of arriving at an estimation of absorbed dose are known as dosemeters or dosimeters. We shall now look further at the four effects noted above and examine how they are used as the basis of working dosemeters.

1. IONISATION OF AIR

Air in its normal state is a good electrical insulator because it contains no conduction electrons. If, however, air is exposed to x- or gamma rays, some of the photons of radiation release electrons from the atoms in the air, ionising it and enabling it to conduct electricity. The more radiation the air is exposed to, the lower its electrical resistance becomes, and the better able it is to conduct electric current. By measuring the electrical properties of the air the quantity of radiation causing the ionisation may be estimated.

Exposure. The measure of the strength of an x- or gamma-ray beam by the quantity of charge on the ions produced in unit mass of air is called exposure. The use of the term 'exposure' is rather different from its more general meaning in radiography as a shortened form of 'exposure factors'.

Roentgen. The unit of radiation exposure is the roentgen (R) defined as the x- or gamma ray exposure which produces a total positive or negative ion charge of 2.58×10^{-4} coulombs per kilogram of dry air. The rather awkward number involved (2.58×10^{-4}) is the result of the conversion to SI units of an older definition. The roentgen is not an SI unit and is being replaced by the coulomb per kilogram (C/kg), where an exposure of 1 C/kg = 2.58×10^{-4} roentgens.

It is possible to convert an exposure measurement into the value of absorbed dose which it represents by the use of a conversion factor; e.g.

Absorbed dose (in rads)
= exposure (in roentgens) × conversion factor

However, this is not quite as straightforward as it seems because the value of the conversion factor is different for different materials. The value may also vary for the *same* material at different beam energies.

The dosemeters which employ the air ionisation effect of radiation use an ionisation chamber as the detection device. We shall look at three types of ionisation chamber dosemeter, the 'free air' ionisation chamber, the 'thimble' chamber, and the quartz fibre electrometer. The first is a laboratory instrument while the other two are used as clinical dosemeters.

FREE AIR IONISATION CHAMBER

Although this instrument is not used in hospitals, it is the device which forms the basis of calibration of hospital dosemeters and it demonstrates more simply than other dosemeters the principles of how ionisation chamber measurements are made. The free air chamber is used to measure radiation exposure. It consists of a box of air. A known mass of air is exposed to a beam of x- or gamma rays and the negative ions produced in the air are collected on a positively charged metal plate. The total charge collected is measured with an instrument called an electrometer. The amount of charge in coulombs is therefore known, and the mass of air from which it was collected is also known. The value of charge/mass can be calculated, giving a direct measure of exposure in coulombs per kilogram or in roentgens if required.

Fig. 18/1 shows the basic design of the free air chamber. Some of the features on the diagram require further explanation:

1. The lead diaphragm restricts the size and shape of the radiation beam passing through the chamber. The dimensions of the beam must be known so that the collecting volume ABCD and therefore the mass of air ionised can be calculated.

2. The electrically charged plates are sufficiently far apart to allow the electrons which are released by the ionising effect of the

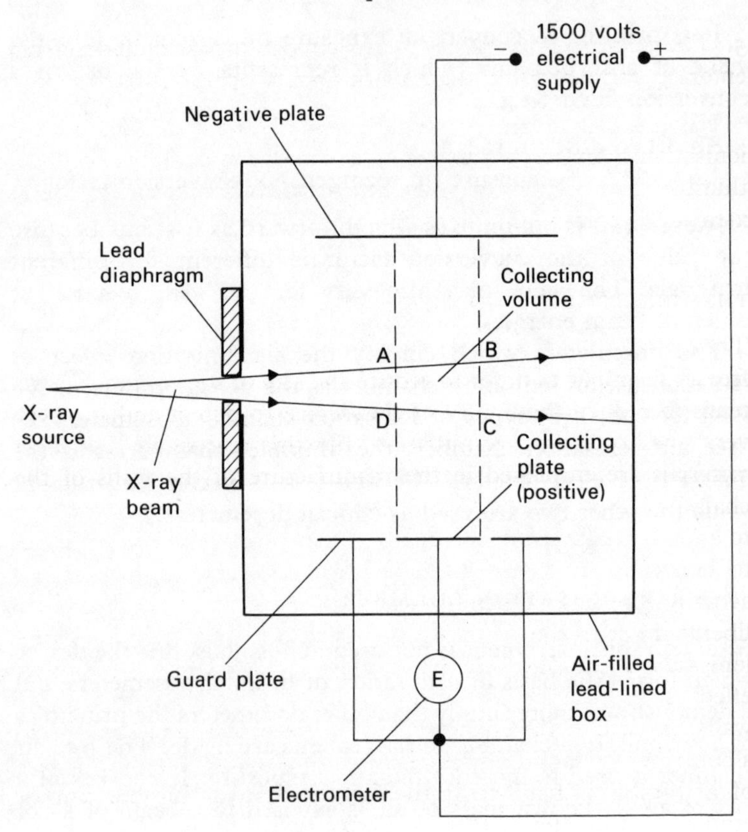

Fig. 18/1. The free-air ionisation chamber.

radiation, to undergo collisions with other atoms, increasing to a maximum the total charge produced, before eventually coming to rest.

3. The charged plates have a sufficiently high potential difference applied to them to ensure that all the electrons released in the air are collected before they have a chance to recombine with any atoms. Only the negative ions (electrons) are collected for measurement, the positive ions being too heavy to be drawn to the collecting plate.

4. One of the charged plates is constructed with a 'guard plate' around it at the same potential. This prevents any distortion of the electric field at the edges of the collecting volume, making it easier to calculate its exact volume.

5. The chamber is lead-lined to prevent extraneous radiation from entering and distorting the measurements being made.

Using some of the principles demonstrated in the free air ionisation chamber, a much smaller ionisation chamber, called a thimble chamber, has been designed, which is suitable for use in x-ray and radiotherapy departments.

THIMBLE IONISATION CHAMBER

The thimble chamber uses a much smaller volume of air than the free air chamber in order to reduce the size of the chamber. This tends to reduce the ability of the device to detect and measure very small quantities of radiation. For this reason *air equivalent* materials are employed in the manufacture of the walls of the chamber. These materials have a similar effective atomic number to air but are much more dense. Thus they allow the chamber to behave as if it had a much greater volume of air than is actually present. When exposed to radiation, electrons are liberated from the walls of the chamber and help to cause ionisation of the air inside the chamber so magnifying the ionising effect of the radiation. Because of the use of air equivalent chamber walls, the size of the chamber can be dramatically reduced from a cumbersome laboratory instrument to a small chamber the size of a thimble. Chambers rather larger than a thimble can be employed if greater sensitivity is required.

Fig. 18/2 shows the construction of a typical thimble chamber. The wall is made of a mixture of bakelite and graphite, or plastic coated with a layer of graphite. Both of these combinations are air equivalent. Graphite is included to make the wall into an

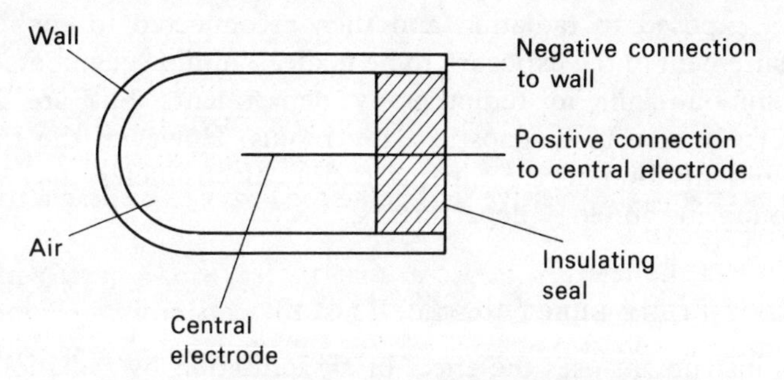

Fig. 18/2. Construction of a thimble chamber.

electrical conductor so that it can be used as part of the electric circuit which is needed to collect the ions liberated by the exposure to radiation.

The central electrode is a thin rod of aluminium. It forms the other part of the ion collecting circuit. The central electrode is made positive and the wall negative, so electrons released inside the chamber are collected on the central rod. The rod is held in position by an insulating seal which closes off the chamber.

The chamber is connected by a cable to the electrical measuring instrument and power supply.

The instrument may be used in two ways:

1. As a dosemeter, it measures the total absorbed dose over a specific time period. Used in this way the charge collected on the central electrode is stored on a capacitor. The amount of charge collected is measured and the total exposure in roentgens (or in coulombs per kilogram) is read off the scale. As we have seen, exposure readings can be converted to absorbed dose by using the conversion factor.

2. As a doserate meter, it measures the absorbed dose produced in unit time. The charge collected per second by the central electrode is found by measuring the electric current flowing in the circuit (remembering that charge/time = current). The exposure rate in roentgens per second can be read off the scale and converted into an absorbed doserate reading in rads per second or grays per second.

Thimble chambers are calibrated by reference to the more basic instrument, the free air chamber.

A modified version of the thimble chamber is available which can be disconnected completely from its electrical circuit and cable, exposed to radiation and then reconnected to enable a measurement of the exposure to be made. Thimble chambers are a common sight in radiotherapy departments but are less frequently seen in diagnostic x-ray rooms. However, the third type of ionisation device, the quartz fibre electrometer, may well be found in radiology departments.

QUARTZ FIBRE ELECTROMETER (QFE)

This instrument uses the effect of air ionisation by radiation in a different way. A very fine fibre of quartz is supported close to a firmly fixed electrode as in Fig. 18/3a. Both are charged with electricity of the same polarity. The electric force of repulsion between the like charges distorts the quartz fibre so that it bends away from the fixed electrode (Fig. 18/3b). When the instrument is exposed to ionising radiation, the air around the fibre and the fixed electrode conducts, partly neutralising their charges and the fibre moves back towards its original position (Fig. 18/3c). The greater the radiation exposure, the further the fibre moves towards the fixed electrode.

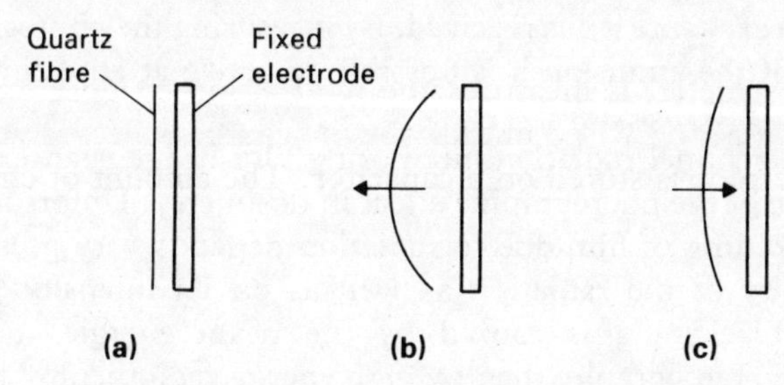

Fig. 18/3. (a) Electrode uncharged; (b) Fibre repelled during charging; (c) Fibre returning during radiation exposure.

A system of lenses is incorporated so that the tiny movements of the quartz fibre can be detected, and a scale is added across which the fibre moves. The QFE is about the size of a fountain pen and can be worn in the pocket as a personnel dosemeter (see Chapter 20). It requires a companion device to provide it with its initial charge before use. While not a very accurate instrument, it has the advantage of being simple to use and a reading of dose is available immediately after exposure to radiation.

We shall now consider the photographic effect, which is also used as a means of measuring radiation doses.

2. PHOTOGRAPHIC EFFECT

The sensitive emulsion on photographic film contains microscopic particles of silver bromide. When this emulsion is irradiated with x- or gamma rays invisible changes occur in the structure of the silver bromide molecules. When the film is developed, the particles of silver bromide which were affected by the radiation are converted into particles of metallic silver. When the film is fixed, any remaining silver bromide is dissolved out of the emulsion. The resulting film therefore contains particles of silver in the parts which have been exposed to radiation, and no silver in unexposed regions. High exposure produces more silver than low exposure.

If the film is held to the light such as from an x-ray illuminator ('viewing box'), light is transmitted by the unexposed parts of the film and absorbed by the exposed parts; i.e. exposed parts appear grey or black. It has been found that the degree of blackening (photographic density) of the film is related to the radiation exposure it has received. By measuring the photographic density of the emulsion it is possible to arrive at an estimate of exposure and dose.

The personnel radiation monitoring film badge is an example of the use of the photographic effect in dosimetry. Unfortunately, the blackening of film due to radiation depends very greatly on the quality of the radiation as well as on its intensity. Much greater blackening is caused by the beam energies used in diagnostic radiography than by high energy radiotherapy beams. A particular degree of blackening, for example, may be caused either by a small exposure to diagnostic x-rays or by a larger exposure to a radiotherapy beam. When using photographic methods of dosimetry, it is therefore necessary to devise some way of estimating the quality of the radiation for which a dose reading is required. In the film badge, plastic and metal filters are incorporated to enable a check to be made on radiation quality. We shall be investigating the film badge dosemeter in more detail when discussing radiation protection in Chapter 20.

3. THERMOLUMINESCENT EFFECT

Certain crystals (e.g. lithium fluoride) are able to store the radiation energy which they absorb when exposed to x-rays. Under the right conditions they can be stimulated to release the stored energy in the form of light some time after the original radiation exposure occurred.

When lithium fluoride is exposed to radiation many of its electrons absorb energy and undergo transitions to orbits further from the nucleus. They remain at these higher energy levels until stimulated by the application of heat. The electrons then fall back to their original shells, while releasing their surplus energy in the form of photons of visible light. So in this process *heat* provides the trigger which prompts the lithium fluoride to release its stored energy. The brightness of the light emitted is related to the absorbed dose, and if the light is measured we can calculate the absorbed dose in rads or grays.

Thermoluminescent dosemeters (TLDs) consist of a small sample of powdered lithium fluoride in a plastic sachet which may be quite tiny. The TLD may be used in many ways; e.g. worn on the person, stuck on the wall or on equipment, or even inserted into a body cavity. When the radiation exposure has stopped, the TLD is taken to a piece of equipment called a 'TLD reader'. Here the lithium fluoride is heated in an electric oven and the light it emits is measured. The instrument is calibrated so that dose may be read off directly from the scale. The lithium fluoride, now in a 'relaxed' state may be re-used many times.

4. FLUORESCENT EFFECT

Fluorescence is an effect similar in some ways to thermoluminescence, but not involving the use of heat. We described the process of fluorescence in Chapter 14 and gave examples of its use in x-ray intensifying screens and television screens.

In dosimetry a crystal of sodium iodide is exposed to radiation. The photons of light it emits are detected by a very sensitive photomultiplier which converts the light impulses into electrical signals and amplifies them thousands of times. The electrical pulses are then counted on an electronic counter. The total pulse count is proportional to the radiation exposure, and the count rate is proportional to the exposure rate. This type of radiation meter is known as a *scintillation counter*.

Up to now, we have considered methods of measuring the exposure and absorbed dose produced by ionising radiations. It is sometimes necessary to consider the quality as well as the quantity of radiation.

EVALUATION OF X-RAY QUALITY

The quality of a heterogeneous x-ray beam is most completely described by reference to its spectrum (see Chapter 15). What is needed is a quicker and simpler method of indicating the main characteristics of a beam on which its quality depends. Three features may be chosen:

1. We may specify the maximum energy of photons in the beam, (or the minimum wavelength). This is easily accomplished with an x-ray beam since the maximum energy photons are produced when electrons reaching the target convert all their energy into x-rays in one interaction. They have a maximum energy which is related to the *peak kilovoltage* (kVp) applied to the x-ray tube, and a minimum wavelength (λ_{min}) which is given by the Duane-Hunt Law:

$$\lambda_{min} = \frac{1.24}{\text{kVp}} \text{ nm.}$$

Thus by quoting the peak kilovoltage applied to the x-ray tube we give a good indication of the quality of the beam. This is the method used in diagnostic radiography.

2. We may specify the effective photon energy of the beam. This is the average energy value of all the photons present in the beam. It represents the photon energy of a homogeneous beam which has the same penetrating ability.

3. We may specify the thickness of filter required to halve the intensity of the beam. This will give us an indication of its ability to penetrate. For example, if one beam needs a copper filter 2 mm thick to halve its intensity, while a second beam needs 6 mm of copper to reduce its intensity by the same fraction, we would conclude that the second beam is more penetrating than the first. The thickness of filter we have defined above is known as the half-value thickness (or half-value layer) of the beam.

HALF-VALUE THICKNESS (HVT)

We define HVT as the thickness of attenuator which, when inserted into a radiation beam, will reduce its intensity by 50%. We can devise a simple practical test to determine the half-value thickness of an x-ray beam. Fig. 18/4 shows the basic requirements for the test.

The x-ray beam is passed through a filter and the intensity of the transmitted beam is measured with a dosemeter. Various thicknesses (t) of filter are inserted and the reading (I) on the

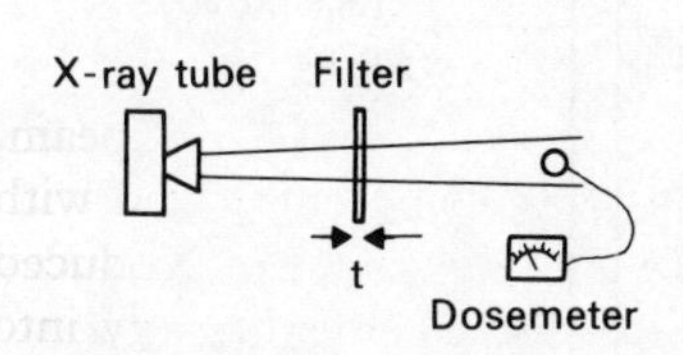

Fig. 18/4. Test to determine the half-value thickness of an x-ray beam.

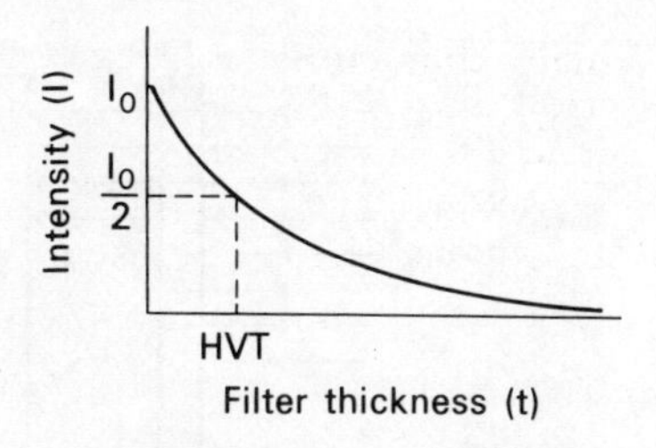

Fig. 18/5. Graph from which half-value thickness can be estimated.

dosemeter recorded in each case. A reading (I_0) is also taken with no filter present. The readings are plotted on a graph (Fig. 18/5).

As we saw in Chapter 16, a graph of transmitted intensity against filter thickness results in an exponential curve. If we check to see what thickness of filter produces a 50% reduction in intensity (i.e. when $I = I_0/2$), we can read off the half-value thickness of the beam. It may prove helpful at this stage if we work through a simple example involving the use of HVT.

PROBLEM

The HVT of a beam of radiation is quoted as 3 mm of tin. What percentage of the beam will be transmitted through a tin filter 9 mm thick?

Remember that if the half-value thickness is inserted, the beam intensity is reduced by 50%. We can consider 9 mm as being three separate layers each of thickness 3 mm. The first 3 mm will reduce the beam to 50%. The second 3 mm will reduce the beam by 50% again, i.e. to 25%. The third 3 mm will reduce the beam by 50% again, i.e. to 12.5%.

We can conclude, therefore, that a 9-mm tin filter will cut the intensity of the beam to 12.5% of its original value. Fig. 18/6 illustrates the method we have used to solve this problem. We have provided the answer in the form of *percentage transmission.* It is also possible to quote the answer as a *fractional transmission* value rather than a percentage. In this case we would say the fractional transmission is 1/8, since 12.5% represents the fraction one-eighth.

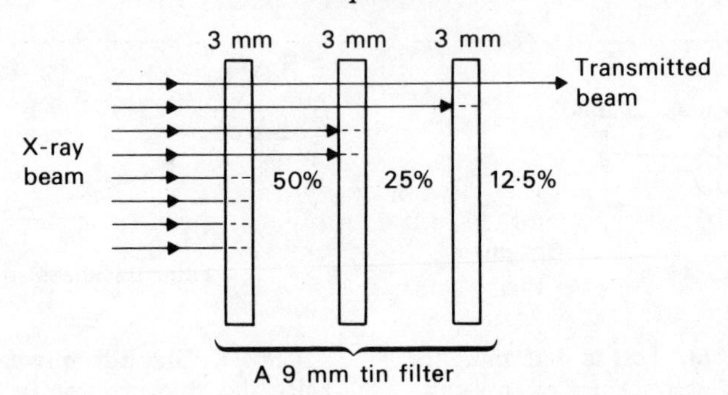

Fig. 18/6. A 9 mm filter, containing three half-value thicknesses, transmitting 12.5% of the incident x-ray beam.

SUMMARY

We have described three possible methods of indicating the quality of a radiation beam. But which of the methods is used in practice? As we pointed out, the peak x-ray tube kilovoltage is generally quoted in diagnostic radiography. In radiotherapy it is usual with beams having energies up to 2 MeV to specify both the generating voltage (or effective photon energy) and the half-value thickness. Above 2 MeV the effective photon energy only need by specified.

In our next chapter we shall examine forms of ionising radiation other than x-rays and describe the radioactive processes which produce them.

CHAPTER SUMMARY

1. Absorbed dose is the energy absorbed per unit mass of the absorbing medium (p. 208).
2. The SI unit of absorbed dose is the gray (p. 208).
3. The value of absorbed dose depends on both beam energy and the type of absorbing medium (p. 208).
4. Exposure is a measure of the ionising effect on air of x- or gamma rays (p. 210).
5. The SI unit of exposure is the coulomb per kilogram. It replaces the unit known as the roentgen (p. 210).
6. The free air chamber is a laboratory instrument used to measure exposure (p. 211).
7. The thimble ionisation chamber is part of a practical dosemeter used in hospitals (p. 213).
8. The quartz fibre electrometer is a pocket dosemeter (p. 214).
9. Monitoring film is used to measure dose by the photographic effect of radiation (p. 216).
10. Lithium fluoride crystals exhibit thermoluminescence and have many applications in dosimetry (p. 216).
11. A scintillation counter employs the fluorescent effect of radiation to measure dose (p. 217).
12. Beam quality may be specified by its maximum photon energy, its effective photon energy or its half value thickness (p. 218).

SOME KEY RELATIONSHIPS

1 gray = 100 rad (p. 208).
Total dose = doserate × time (p. 209).
1 coulomb/kilogram = 2.58×10^{-4} roentgens (p. 210).
Absorbed dose = exposure × conversion factor (p. 211).

$\lambda_{min} = \frac{1.24}{kVp}$ nm (Duane-Hunt Law) (p. 218).

Chapter 19
Radioactivity

(Chapter 4 should be reviewed before studying this section of work on radioactivity.)

In Chapter 4 we said that the nuclei of some nuclides (isotopes) are unstable and they tend to fly apart, emitting radiation and fast moving particles. We called these unstable nuclides *radionuclides* or *radioisotopes* and described them as being *radioactive*.

RADIOACTIVITY

Radioactivity, or radioactive decay, is defined as a process whereby some nuclides undergo spontaneous changes in the structure of their nuclei, accompanied by the emission of particles and radiation.

Radioactivity is a nuclear process, i.e. it involves the nuclei of atoms rather than the electrons in orbit around the nuclei. Thus radioactivity is not influenced by chemical changes that may be occurring, nor by changes in the physical environment such as variations in temperature or pressure. It is therefore not possible to control the rate of radioactive breakdown of a nuclide.

CAUSES OF RADIOACTIVITY

Why do some nuclides disintegrate while others are stable? The nuclei of atoms contain two kinds of fundamental particle: positively charged protons and uncharged neutrons (see Chapter 4). Because the protons carry similar charges, electric forces are set up between them which cause them to repel each other. If unrestrained, the protons would separate causing the nucleus to break up. However, the presence of neutrons in the nucleus counteracts this tendency, and as a result the nucleus may survive intact.

In order for the repulsion forces to be overcome in a particular nucleus, there must be a specific number of neutrons present. The 'neutron : proton ratio' must be correct. If too many or too few neutrons are present, sooner or later nuclear disintegration will take place.

Some nuclides, particularly ones having high atomic numbers, are unstable no matter how many neutrons are present; e.g. uranium, whose atomic number is 92, has no stable isotopes.

After the disintegration of a radionuclide, a new nuclide is formed which may, or may not be stable. The nuclides which result from radioactive transformations are known as *daughter products*. In some instances a whole series of transformations may be involved before a stable daughter nuclide is achieved.

NATURAL AND ARTIFICIAL RADIOISOTOPES

Many unstable isotopes occur naturally in the earth, in living organisms and in the atmosphere; e.g. $^{238}_{92}U$ (uranium), $^{226}_{88}Ra$ (radium), $^{40}_{19}K$ (potassium), and $^{14}_{6}C$ (carbon). Other radioisotopes have been produced by man in nuclear reactors and in high energy machines called particle accelerators; e.g. $^{131}_{53}I$ (iodine), $^{90}_{38}Sr$ (strontium), $^{60}_{27}Co$ (cobalt) and $^{99}_{43}Tc$ (technetium).

TRANSFORMATION PROCESSES

When radionuclides break down, several different types of emission may be produced depending on which particular process of decay has occurred. We shall study just three of the many possible transformation processes.

1. ALPHA (α)-PARTICLE EMISSION

The emission of an alpha particle is associated with the breakdown of heavy elements such as uranium and radium. The alpha particle is a combination of fundamental particles: namely two protons and two neutrons; i.e. an alpha particle is identical in structure to the nucleus of a helium atom ($^{4}_{2}He$). The decay of radium is an example of α-emission:

$$^{226}_{88}\text{Ra (radium)} \rightarrow {}^{222}_{86}\text{Rn (radon)} + {}^{4}_{2}\alpha \text{ (alpha particle)}$$

The alpha particle is ejected at high speed from the nucleus. Notice that both the atomic numbers and the mass numbers balance on both sides of the equation. Notice also that in this process the mass number of the nuclide has reduced by *four* and the atomic number has reduced by *two*.

The daughter product, radon (a gas), is also radioactive and undergoes a further alpha-particle emission. After a total of nine successive transformations a stable nuclide is formed, one of the isotopes of lead.

Properties of alpha particles. Alpha particles carry a positive electric charge and cause intense ionisation in matter through which they pass. They are the heaviest of the particle emissions. Because they are electrically charged, their course is influenced by both electric and magnetic fields; e.g. they would be attracted towards a negatively charged electrode.

Alpha particles have a very short range. Even in air their energy will become attenuated after travelling only a few centimetres. A sheet of paper would stop them completely. The effect of alpha particles on living tissues is intense, but very localised.

2. BETA (β)-PARTICLE EMISSION

The emission of a beta particle is associated with the breakdown of nuclides which have too many neutrons; i.e. nuclides whose neutron : proton ratio is too high. The beta particle is identical to an electron, having the same mass and carrying the same charge. It results from the transformation of a neutron (n) in the nucleus into a proton (p); i.e.

$$^{1}_{0}\mathrm{n} \rightarrow {}^{1}_{+1}\mathrm{p} + {}^{0}_{-1}\beta$$

The beta particle is ejected from the nucleus at high speed. As a result of the transformation the number of neutrons in the nucleus is reduced by one, while the number of protons is increased by one, thus decreasing the neutron : proton ratio to a more acceptable level. The breakdown of radioactive caesium ($^{137}_{55}\mathrm{Cs}$) is an example of β-emission:

$$^{137}_{55}\mathrm{Cs}\ \text{(caesium)} \rightarrow {}^{137}_{56}\mathrm{Ba}\ \text{(barium)} + {}^{0}_{-1}\beta$$

Notice that in this process the mass number remains the same, while the atomic number increases by one.

Properties of beta particles. Beta particles carry a negative electric charge and cause ionisation in any medium they traverse. They can be deflected by electric and magnetic fields, but in the opposite sense to alpha particles; e.g. they would be repelled away from a negatively charged electrode.

Beta particles have a greater range than alpha particles and would not be much attenuated by a sheet of paper, but would be stopped by a layer of aluminium a few millimetres thick. Beta particles, too, have a localised damaging effect on body tissues.

We have described negatively charged beta particles (β^-), but positive beta particles (β^+) can also be emitted through a rather different process which we shall not describe. It is, however, important to be aware of the existence of β^+ particles and not to confuse them with the β^- variety.

3. GAMMA-RAY EMISSION

Alpha- and beta-decay processes usually leave the daughter nucleus in an excited state, having too much energy. The nucleus sheds its surplus energy by emitting a photon of high energy electromagnetic radiation called gamma (γ) radiation. The isotope of barium ($^{137}_{56}Ba$), which was produced in our example of beta particle emission, behaves in this way:

$$^{137}_{56}\text{Ba}^{\star} \rightarrow {}^{137}_{56}\text{Ba} + \gamma$$

The asterisk is used to indicate that a nuclide is in an excited state. Notice that unlike alpha and beta decay, gamma-ray emission involves no alteration in the atomic number or the mass number.

The energy of gamma-ray emission is always the same for a particular nuclide; i.e. it is characteristic of a specific transformation, and produces a line spectrum rather than characteristic spectrum (Chapter 14).

Properties of gamma rays. Gamma rays have no mass and carry no electric charge, being radiation rather than matter. They are therefore not influenced by electric or magnetic fields but they do have an ionising effect on matter, interacting with it in the same way as x-rays (see Chapters 16 and 17).

Gamma rays are far more penetrating than alpha or beta particles, requiring the equivalent of several millimetres of lead to produce significant attenuation. Gamma rays are able to penetrate into body tissues and are therefore used in the radiotherapy of deep-lying lesions. Radioactive cobalt ($^{60}_{27}Co$) is a common source of gamma rays for this purpose.

Having now completed our description of the main transformation processes and the emissions they produce, we shall consider the rate at which these processes occur and the means we have of measuring radioactivity.

RADIOACTIVE DECAY RATES

Radioactivity is a random process. We can never predict when a particular atomic nucleus will disintegrate. However, we are able to make predictions about the behaviour of large numbers of atoms such as we deal with in practice. (In a similar way, we are not able to predict whether a particular expectant mother will give birth to a boy or a girl, but we can say with some confidence that in a group of 100 expectant mothers, approximately half will have boys and half girls.)

The rate of decay of a radionuclide is known as its *activity*. The activity of a sample is the rate at which the sample undergoes transformations, i.e. the number of transformations per second.

One gram of radium has an activity of 3.7×10^{10} transformations per second and this has traditionally been used as the unit of activity. The unit is named the *curie* (Ci) and is defined as an activity of 3.7×10^{10} transformations per second. This is a relatively high activity. For smaller values we can use the *millicurie* (mCi) or the *microcurie* (μCi), where:

$$1 \text{ mCi} = 10^{-3} \text{ Ci}$$
$$\text{and} \quad 1 \ \mu\text{Ci} = 10^{-6} \text{ Ci}.$$

The curie is now in the process of being replaced by the SI unit of activity known as the *becquerel* (Bq). This is a much smaller unit, being defined as an activity of *one* transformation per second.

$$\text{Thus} \quad 1 \text{ Ci} = 3.7 \times 10^{10} \text{ Bq}$$
$$\text{and} \quad 1 \text{ Ci} = 3.7 \times 10^{4} \text{ megabecquerels}$$

where $1 \text{ MBq} = 10^6$ Bq.

A cobalt gamma-ray source used in radiotherapy has an activity of up to 5000 Ci (2×10^8 MBq). A luminous wristwatch face has an activity of about 10 μCi (0.4 MBq).

As a radioactive sample decays, its activity reduces because as time passes, there is less and less active material left. If we plot a graph of the activity of a sample against time (Fig. 19/1) we obtain the familiar exponential decay curve. Its mathematical relationship is in the form:

$$A = A_0 e^{-\lambda t}$$

where: A is the activity after time t, A_0 is the initial activity of the sample (i.e. when $t = 0$), e is the exponential constant ($e = 2.72$, λ is called the *decay constant* and its value is constant for a particular radionuclide.

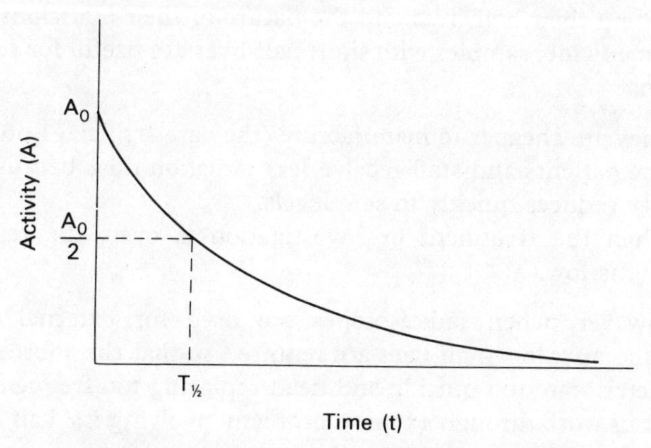

Fig. 19/1. The decay of a radioactive sample.

RADIOACTIVE HALF LIFE

Each radionuclide has its own decay rate which we can specify either by means of the decay constant, or more simply by noting the time taken for the activity of the radionuclide to decrease by 50%. We call this time the *radioactive half life* ($T_{\frac{1}{2}}$). It is indicated on the graph in Fig. 19/1.

The value of half life for different radionuclides varies from minute fractions of a second (e.g. 10^{-9} seconds) to over a million years (e.g. 10^{10} years).

The decay constant and half life are related by the equation:

$$\lambda = \frac{0.693}{T_{\frac{1}{2}}}.$$

Table 19/1. Examples of radionuclides used in medicine.

Radionuclide	Half life	Emission(s)	Medical use
Radium 226	1620 years	Alpha and gamma	Therapy source
Cobalt 60	5.24 years	$Beta^-$ and gamma	Therapy source
Caesium 137	33 years	$Beta^-$	Therapy source
Iodine 131	8.1 days	$Beta^-$ and gamma	Scanning
Gold 198	2.7 days	$Beta^-$ and gamma	Therapy source
Technetium 99	6 hours	Gamma	Scanning
Oxygen 15	126 seconds	$Beta^+$	Oxygen uptake

Table 19/1 shows some of the radionuclides used in hospitals, indicating their half lives and the nature of their emissions.

In medicine, samples with short half lives are useful for several reasons:

1. They are cheaper to manufacture (they are artificial isotopes).
2. The patients and staff receive less radiation dose because the activity reduces quickly to safe levels.
3. When the treatment or investigation is over, the residual activity is low.

However, when radioisotopes are used for external beam radiotherapy, long half lives are required so that the source does not deteriorate too quickly and need replacing too frequently.

Let us work through a typical problem involving the half life of a radioisotope.

PROBLEM

A sample of iodine 131 with a half life of 8 days is to be used for an isotope scanning procedure. The delivery of the sample from the supplier takes 16 days. If an activity of 1 MBq is required to carry out the investigation, what should the activity of the sample be when it is despatched from the supplier?

Remember that each 8 days reduces the activity by 50%. Eight days prior to the investigation, therefore, the activity of the sample would have been twice its final activity; i.e. 8 days before use the activity was $2 \times 1 = 2$ MBq. Eight days before that (i.e. 16 days before use) the activity would have been doubled again; i.e. 16 days before use the activity was $2 \times 2 = 4$ MBq. So when the sample leaves the supplier, its activity should be 4 MBq.

To carry out some investigations or treatments, we may need to know what *mass* of a radioactive sample will give a particular activity level. We need to know the *specific activity* of the sample, which is defined as the mass whose activity is 1 Ci (or 1 Bq in SI units). It may be quoted in curies per kilogram or becquerels per kilogram.

In our final chapter we consider the *biological* effects of ionising radiation and discuss the measures which may be taken to minimise any hazards.

CHAPTER SUMMARY

1. Radioactivity is the process of spontaneous transformation of atomic nuclei, accompanied by the emission of particles and radiation (p. 222).
2. Alpha-particle emission is associated with breakdown of heavy elements (p. 223).
3. Beta-particle emission is associated with the breakdown of nuclides with too many neutrons (p. 224).
4. Gamma rays are emitted by daughter products with surplus energy (p. 225).
5. The activity of a radionuclide is the number of transformations occurring per second. The SI unit of activity is the becquerel (p. 226).
6. Radioactive half-life is the time taken for the activity of a radionuclide to decrease by 50% (p. 227).

SOME KEY RELATIONSHIPS

1 curie $= 3.7 \times 10^{10}$ Bq (p. 226).

$A = A_0 e^{-\lambda t}$ (radioactive decay) (p. 227).

$\lambda = \dfrac{0.693}{T_{\frac{1}{2}}}$ (decay constant) (p. 227).

Chapter 20
Radiation protection

This chapter should be studied in conjunction with the *Code of Practice for the Protection of Persons against Ionizing Radiation arising from Medical and Dental Use*, published by Her Majesty's Stationery Office.

In this chapter we discuss the harmful effects of ionising radiations on the human body, and see what can be done to minimise the damage such radiations cause.

HISTORICAL BACKGROUND

In the years immediately after the discovery of x-rays and radioactivity, many of the pioneering workers were seriously affected by exposure to radiation because they were not aware of the potential dangers and they had no knowledge of what levels of radiation were likely to cause damage. In many cases visible signs of damage did not appear until perhaps 20 years after the exposure.

Workers who handled radium developed malignant growths on their hands, often leading to amputation. Any radium which was accidentally ingested was laid down like calcium and strontium in the bony skeleton, resulting in leukaemia.

Eventually 'safe' levels were devised which it was thought would not lead to such damage. Over the years these levels have been gradually reduced because the levels which were originally thought to be safe have since been found to cause long term effects which are not acceptable. The permitted levels are now set down in the recommendations of the International Commission for Radiation Protection (ICRP). Each nation interprets the recommendations and produces a Code of Practice for radiation workers who must read the part of the code relevant to their particular job.

UNITS OF MEASUREMENT

Different forms of radiation tend to produce different degrees of biological damage even for the same absorbed dose. When discussing radiation measurements it is therefore useful to modify the absorbed dose value to take this into account. The modification is achieved by means of a term known as the *quality factor* of the radiation. The modified form of dose measurement is called the *dose equivalent*, obtained from the relationship:

$$\text{Dose equivalent} = \text{absorbed dose} \times \text{quality factor.}$$

The unit of dose equivalent is the *rem*, where:

$$\text{Dose equivalent (in rems)} = \text{absorbed dose (in rads)} \times \text{quality factor.}$$

The rem is not an SI unit. (The SI unit is the sievert (Sv) and is replacing the rem as the unit of dose equivalent: 1 Sv = 100 rems.)

Conveniently, for x-rays and gamma rays the quality factor is one. Thus for these particular radiations the dose equivalent in rems is numerically equal to the absorbed dose in rads. Note that for small doses the millirem (mrem) is used, where 1 millirem = 10^{-3} rem.

BIOLOGICAL EFFECTS OF RADIATION

Ionising radiations can damage all living tissues to some extent. If the damage is limited to a small number of comparatively unimportant cells, the total effect on the body may be insignificant. But if a large number of cells, or even one cell vital to the functioning of the body, is damaged, then this may constitute a hazard to health.

We classify biological effects according to whether they are associated with the general tissues of the body (somatic effects) or the reproductive cells in particular (genetic effects).

1. Somatic effects. These are subdivided into short term and long term effects.

Short term effects arise soon after the exposure to radiation and may subside quite quickly; e.g. 1000 rems of x-rays received over a small area of skin produces reddening (erythema) rather like sunburn, but the effect fades rapidly. The same dose is fatal if received over the whole body because the volume of tissue irradiated is critical.

Long term effects only become apparent years after the exposure to radiation occurred. Local effects include the formation of skin tumours and cataracts on the eye. If the whole body is irradiated leukaemia may develop.

2. Genetic effects. These affect the population as a whole rather than individuals. They are caused by radiation damage to the hereditary material (genes) passed on from parents to their children and are the consequence of irradiation of the male or female gonads. The genetic mutations resulting may be beneficial, but are very much more likely to prove harmful; e.g. congenital blindness, deafness, mental and physical abnormalities and in many cases fetal death. Such mutations occur naturally but excessive exposure to radiation increases the risk. It is mainly for genetic reasons that permitted levels of radiation are set so low.

MAXIMUM PERMISSIBLE DOSE OF RADIATION (MPD)

A maximum permissible dose is not an absolute level of safety, below which we are safe and above which we are unsafe. The MPD is defined statistically to limit the genetic mutation rate of the general population to acceptable levels. The MPD is a *mild* dose, unlikely to cause somatic effects, and designated workers, special categories of people such as radiographers who work with radiation for long periods, are required to accept a larger dose than the average for the population as a whole. The general population is therefore set a correspondingly lower MPD to maintain the overall mutation rate at an acceptable level.

Definition. The maximum permissible dose is defined as:

1. The dose of ionising radiation which would not cause damage that a person would consider objectionable or competent medical authorities consider a danger to health.

2. The dose that a designated person working a 40-hour week could receive indefinitely without causing noticeable damage.

Note that MPDs do not apply to doses received by patients as part of the diagnosis or treatment of disease. In these cases, the benefits from the use of radiation probably far outweigh the risks involved.

EXAMPLES OF MAXIMUM PERMISSIBLE DOSE

DESIGNATED PERSONS

1. For a designated worker such as a radiographer the annual MPD is 5 rems. This is equivalent to a rate of about 100 mrems per week (or 2.5 mrems per hour, based on a 40-hour working week).
2. It is also stipulated that no more than 3 rems may be received in any quarter of the year (13 weeks).
3. The accumulated dose up to the age of 30 should not exceed 60 rems.

The maximum permitted total dose (D) received up to the age of N years is given by:

$$D = (N - 18)5 \text{ rems},$$

i.e. 5 rems for each year of age over 18 years. So a 28-year-old radiographer has a maximum permitted accumulated dose of

$$D = (28 - 18) \times 5 = 50 \text{ rems}.$$

GENERAL PUBLIC

The MPD for the general public is one-tenth of that for a designated worker; i.e. 0.5 rem per year.

The MPD values we have been describing are doses received by the gonads or by the whole body i.e. bone marrow, reproductive organs and the cornea of the eye. These are regions of the body which are highly sensitive to ionising radiation. The MPD values for the skin, thyroid gland and bone are higher, because these regions are of medium sensitivity. While the MPDs for the extremities (hands, forearms, feet and lower legs) are higher still. The Code of Practice sets out the maximum permissible doses in detail and should be studied by all radiographers.

NATURAL BACKGROUND RADIATION

All individuals are exposed to low levels of ionising radiation from natural sources such as:

1. Cosmic radiation from outer space.
2. Radiation from radioactive elements in the earth's crust and in the atmosphere.
3. Radiation from radioactive elements in the body.

The annual dose to the gonads per person from these sources amounts to about 100 mrem. This gonad dose is unavoidable, but the population is also exposed to ionising radiation from artificial sources such as television sets, luminous paint, occupational use of radioisotopes, nuclear test fall-out, and from medical uses of radiation. The annual gonad dose per person from these sources amounts to about 22 mrem. Almost 90% of this total is due to radiography (5 mrems per year from radiotherapy and 14 mrems per year from diagnostic radiography). As radiographers, we are responsible for administering the greater part of the radiation dose received by the general population over and above that received from natural sources. This is why a strict Code of Practice is so important.

SUMMARY OF THE CODE OF PRACTICE

The Code of Practice:
(a) sets out the basic principles of radiation protection;
(b) gives general guidance on good practice;
(c) deals with radiation hazards to patients;
(d) deals with the disposal of radioactive waste.
It defines four categories of persons:

1. *Occupationally exposed persons.*
   (a) Designated. Those whose doses might exceed 30% of the annual maximum permitted dose. Designated persons must have their radiation doses monitored and must be subject to medical supervision. Radiographers are designated persons.
   (b) Other persons. Those whose doses are unlikely to exceed

30% of the annual MPD. This group is not individually monitored although the working environment should be monitored.

(*N.B.* It is arguable whether radiographers should be classed as designated persons or as other persons, since it is rare for a radiographer to receive a dose exceeding 30% of the annual MPD. Discuss this point with your tutors and with colleagues.)

2. *Patients.* Those who are undergoing examination or treatment by ionising radiations.

3. *Members of the Public.* This group includes other patients who are not in category (2) above.

4. Persons subjected to investigations involving radiation for research purposes.

The Code of Practice sets out an administrative chain of responsibility. The responsibility for implementing the Code of Practice lies ultimately with the Controlling Authority (e.g. the Area Health Authority) who set up a Radiological Safety Committee which considers the reports of a Radiological Protection Adviser (usually a physicist).

In each department where radiation is used the head of the department is responsible for observance of the Code of Practice. The head of department appoints a Radiological Safety Officer (often a senior or superintendent radiographer) to ensure adequate protection measures are carried out, and a Supervisory Medical Officer responsible for the medical supervision of all staff concerned. Local rules must be made which supplement the Code of Practice and set out in detail local procedures and the names and duties of the officers noted above.

The Controlling Authority must see that all staff working with radiation are adequately instructed about hazards and precautions and that each designated person reads the relevant sections of the Code and the local rules and signs a statement confirming that he is aware of and understands these responsibilities.

Medical examinations. These are carried out on designated staff under the authority of the Supervisory Medical Officer. The purpose of the examinations is:

1. To assess medical fitness to perform the task properly;

2. To set a 'base-level' record from which to judge possible future changes in health. A full blood count is part of this initial examination.

3. To monitor continuing fitness.

As long as radiographers consistently receive less than 30% of the annual MPD no annual medical examination is necessary.

Personnel monitoring. Each designated person should wear a personnel dosemeter at chest or waist level. It should be worn underneath any protective apron which may be used. Individual dose records must be kept, and when leaving employment, radiation dose transfer records must be available giving details of the accumulated dose received and any overdoses recorded.

Environmental monitoring. The radiation levels in x-ray and radiotherapy departments must be surveyed and recorded at regular intervals. This is particularly important where x-ray equipment has been replaced or the work load significantly altered.

PROTECTIVE MATERIALS

Before we discuss the radiation safety aspects of departmental design we need to examine the features desirable in protective materials. Lead is the most widely known protective material. It has a high atomic number ($Z = 82$) and high density. At the lower radiation energies employed in diagnostic radiography we know that beam attenuation is due mainly to photoelectric absorption and is therefore strongly influenced by the atomic number of the attenuating medium. In many ways lead is therefore an ideal protective material for diagnostic radiography. For some purposes, however, lead lacks sufficient structural strength and cannot be used alone as a building material for walls, ceilings, etc. In these circumstances we use normal building materials such as brick or concrete. However we still use lead as the standard with which to compare these materials, referring to their *lead equivalent* value.

LEAD EQUIVALENT

The lead equivalent of a protective barrier is the thickness of lead which offers the same degree of protection at a particular radiation quality. For example, a concrete wall 15 cm thick offers the same protection from 100-keV x-rays as a layer of lead 2.4 mm thick. We therefore say that its lead equivalent is 2.4 mm at 100 keV.

At higher radiation energies photoelectric absorption is much less important, Compton scatter being the main attenuation process. Thus the atomic number of protective materials is less critical with high energy x-rays and gamma rays, and materials such as concrete compare more favourably with lead; e.g. at 3 MeV the lead equivalent of 15 cm of concrete is 26 mm. In radiotherapy departments it is feasible to construct the walls of treatment rooms with concrete which possesses the required structural strength. Better protection can be provided if necessary either by the addition of barium sulphate to the concrete mixture (known as *barium concrete*) or by coating the walls with a layer of plaster into which barium has been mixed (known as *barium plaster*).

Lead is used as a ray-proof lining inside x-ray tube housings and treatment heads for cobalt units. It is sandwiched between layers of wood to form lead plywood and used in the manufacture of ray-proof doors, mobile barriers, control cubicle surrounds, etc.

Lead salts can be added during the manufacture of glass to produce the lead glass used for ray-proof viewing windows. Lead salts are also incorporated during the production of flexible material known as lead-rubber, used for protective aprons and gloves. Typically a lead-rubber apron has a lead equivalent of 0.25 mm and provides protection from scattered radiation only.

PROTECTION IN X-RAY AND RADIOTHERAPY DEPARTMENTS

The aim of protective measures is to shield members of staff and patients from the primary radiation beam and from scattered or secondary radiation by limiting the dose received by these individuals to well below their maximum permissible doses.

DIAGNOSTIC DEPARTMENTS

The protection of staff is relatively simple in radiography because the photon energies employed are low (up to 150 keV), and exposure times are short. A barrier of 1 mm lead equivalent provides adequate protection from scatter; e.g. the control panel may be positioned behind such a barrier and some distance from the source so that the radiation is attenuated by distance (Inverse Square Law) as well as by the barrier. The patient is the main source of scattered radiation so we must ensure that during the exposure we position ourselves so that there is a barrier between us and the patient. The primary beam is not such a hazard as its direction is known and it should not be pointed at areas occupied by staff.

The dose received by the patient is minimised by care and attention to such points as:

1. Accurate collimation of the primary beam (coning down).
2. The use of fast or rare earth intensifying screens where appropriate to reduce exposure factors.
3. The use of recommended tube filtration (2.5 mm of aluminium for 100–150 kVp exposures).
4. Adequate immobilisation and accurate positioning of the patient to avoid the need for repeat exposures.
5. The use of gonad protectors in appropriate cases.

N.B. The irradiation of undiagnosed early pregnancies is avoided by adherence to the *Ten-Day Rule* which states that female patients of child-bearing age should have radiographic examinations of the abdomen and pelvis only during the first 10 days after onset of menstruation. During this time the possibility of pregnancy is at its lowest, and the risk of irradiating an immature fetus is minimised.

There are other methods of reducing patient dose which you should discuss with your tutor and your colleagues.

Fluoroscopic (screening) examinations are potentially more hazardous because the radiologist may have to stand in direct line with the primary beam, and the radiographer assisting him may have to stand very close to the patient. The radiologist is protected by a primary barrier (a lead-glass screen or the ray-proof housing around the image intensifier), which is permanently mounted on the x-ray unit and which follows the movements of the under-couch x-ray tube. He is also protected by a lead-rubber curtain and by wearing a protective apron and possibly lead-rubber gloves. Some of these features are illustrated in Fig. 20/1. The radiographer is adequately shielded by a protective apron and by standing well back from the patient whenever possible. Modern fluoroscopy units may have a remote control facility by

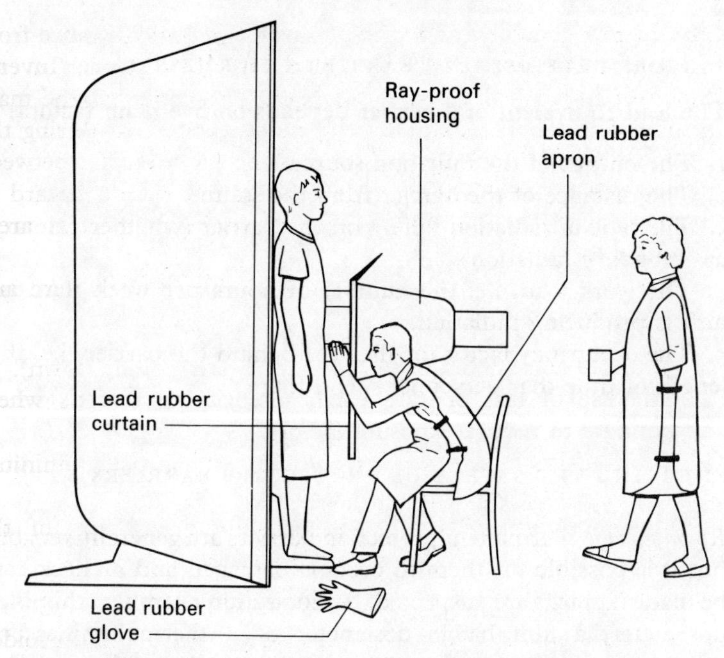

Fig. 20/1. Protective materials during fluoroscopy.

which the set may be operated from behind a permanent protective barrier.

The patient is protected during fluoroscopy by adherence to the points relating to radiographic exposures which we have already discussed, and also by (a) the use of image intensifiers to reduce the tube current required; (b) the use of fluoroscopy timing devices which remind the radiologist of the length of time he has been screening each patient.

THERAPEUTIC DEPARTMENTS

With the exception of low energy superficial therapy, staff are protected by being excluded from the treatment room while treatment is in progress. Interlocks prevent entry into the room unless the radiation is cut off. Any wall, floor or ceiling at which the primary beam may be pointed must be designed as a primary barrier. Surfaces at which the beam cannot be directed need only be secondary radiation barriers. It is economically desirable to restrict to a minimum the directions in which the beam may be pointed in order to limit the area of primary shielding required.

FACTORS DETERMINING BARRIER THICKNESS

The lead equivalent of a barrier depends on five main factors:

1. The output of the radiation source.
2. The distance of the barrier from the source.
3. The type of radiation falling on the barrier (whether primary or secondary radiation).
4. The work load, i.e. the number of hours per week that the unit is producing radiation.
5. The occupancy factor for the area behind the barrier, i.e. the length of time that personnel spend there.

CHECKING THE SAFETY OF PROTECTIVE BARRIERS

Room surveys. Permanent protection barriers are generally reliable but it is possible for them to become damaged and a check can be made if cracks are suspected. An ionisation chamber (thimble) dosemeter, a film badge dosemeter or a thermoluminescent dosemeter may be employed to monitor radiation levels. The film dosemeters could be left in position for a week or more to take into account the departmental work load.

Protective clothing. Flexible lead-rubber deteriorates with wear and should be checked regularly for safety. This can be done either (a) by taking radiographs of the suspect material (cracks would show up as fine black lines on the processed radiograph); or (b) by examining the material fluoroscopically with an image intensifier (cracks appear as fine white lines on the television monitor).

PERSONNEL MONITORING

The Code of Practice stipulates that designated staff must have the doses they receive at work continuously monitored. To achieve this a small pocket dosemeter is required. For many years the traditional method has been to employ the photographic film badge which we described briefly in Chapter 18.

FILM BADGE DOSEMETER

You will remember from Chapter 18 that the degree of blackening of film is roughly proportional to the radiation exposure it receives, but that blackening is also very dependent on radiation quality. Films are much more sensitive to diagnostic x-rays than to higher energy beams, so to obtain dose readings from the film a method of estimating radiation quality is incorporated into the film holder.

The sensitivity of film varies slightly from one batch of film to another and is also affected by processing conditions. These factors must therefore be standardised if accurate measurements are to be made.

Advantages of the film dosemeter. It is robust, provides a permanent record of dose, and is able to measure a wide range of doses (from 10 mrem to 20 rem).

The extended measurement range is achieved by using a film with two emulsions; a slow emulsion coated on one side of the base and a faster emulsion on the other side. If a high dose is received the fast emulsion which after processing is completely black is stripped off, leaving the slow emulsion with a useable degree of blackening.

Filters. The filters incorporated into the film holder are arranged as shown in Fig. 20/2, but the best way to study the design of the film holder is to open your own personnel dosemeter and examine its construction. The higher the radiation energy to which

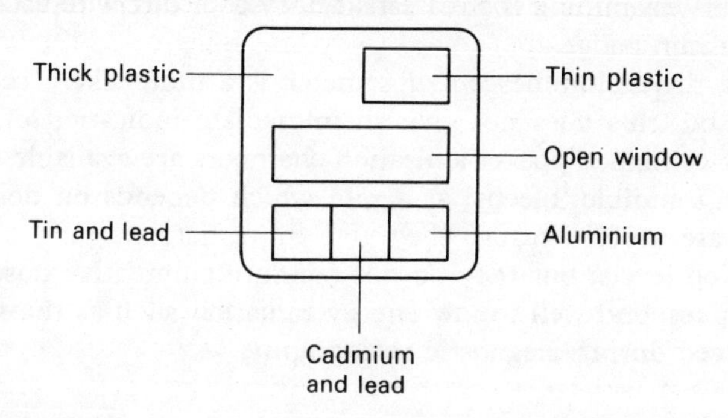

Fig. 20/2. Filters incorporated in the film badge holder of a personal dosemeter.

the badge is exposed, the more uniform will be the penetration through the various filters. By measuring the differences in blackening between them an estimate can be made of the radiation quality.

HOW THE FILM DOSEMETER IS USED

For each film sent out to a hospital from the Radiological Protection Service (RPS), a companion film chosen from the same original batch of film is kept in store as a control. The films are embossed with an identification number which is allocated to the wearer of the film. On receipt at the hospital, the film is issued to the appropriate designated worker who wears it in his plastic film holder for a period of 2 weeks (or 1 month). The film is then returned to the RPS and matched with its control companion. The two films are processed together. The control film is used to estimate the sensitivity of the film while the blackening of its companion is matched against film which has received known exposures. The dose estimates are recorded and copies sent back to the hospital.

The main disadvantage of the film dosemeter system is the delay between an overexposure occurring and the detection of the overexposure on the film. The delay could be as long as 5 weeks.

OTHER METHODS OF PERSONNEL DOSIMETRY

The quartz fibre electrometer dosemeter described in Chapter 18 is a possible alternative that gives an immediate dose reading, but in other ways it is less than satisfactory as a direct replacement for the film badge.

The thermoluminescent dosemeter is a more likely replacement, but this does not give an immediate indication of dose.

Self-contained pocket ionisation chambers are available which give out audible 'bleeps' at a rate which depends on doserate. They are useful because they give immediate warning of high radiation levels, but they do not measure cumulative dose, and do not respond well to low energy radiation such as the scatter produced during diagnostic radiography.

CHAPTER SUMMARY

1. Ionising radiations are potentially harmful to all living tissues (p. 231).
2. Biological damage is classified as being either a somatic or a genetic effect of radiation (p. 231).
3. The dose equivalent of radiation is a measure of its effect on living tissues; it is measured in rems (p. 231).
4. Maximum permissible doses are mild doses unlikely to cause somatic effects in individuals or significantly raise the genetic mutation rate in the general population (p. 232).
5. The annual MPD is 5 rems for designated workers and 0.5 rems for the general public (p. 233).
6. The Code of Practice sets out the basic principles of radiation protection (p. 234).
7. Lead equivalent of an attenuating barrier is the thickness of lead which offers the same degree of protection at a particular radiation quality (p. 236).
8. The film badge dosemeter is the most common method of personnel dosimetry (p. 241).

KEY RELATIONSHIP

Dose equivalent = absorbed dose × quality factor (p. 231).

Suggested further reading

These books offer a more detailed description of general and radiation physics:

JAUNDRELL-THOMPSON F. & ASHWORTH W. J. (1970) *X-Ray Physics and Equipment*. Blackwell Scientific Publications, Oxford.

MEREDITH W. J. & MASSEY J. B. (1977) *Fundamental Physics of Radiology*. John Wright, Bristol.

RIDGWAY A. & THUMM W. (1968) *The Physics of Medical Radiography*. Addison-Wesley, London.

These books cover a few aspects of a specialised nature:

HER MAJESTY'S STATIONERY OFFICE (1972) *Code of Practice for the Protection of Persons against Ionizing Radiations arising from Medical and Dental Use*. HMSO, London.

HILL D. R. (1975) *Principles of Diagnostic X-ray Apparatus*. MacMillan, London.

I.C.R.P. (1970) *Protection of the Patient in X-ray Diagnosis*. Publ. No. 16. Pergamon, Oxford.

OLIVER R. (1966) *Radiation Physics in Radiology*. Blackwell Scientific Publications, Oxford.

WORLD HEALTH ORGANISATION (1977) *The S.I. for the Health Professions*. WHO, Geneva.

Index

3